131 P - 69

D1200884

ELEMENTS OF
CLOUD PHYSICS

64238

ELEMENTS OF CLOUD PHYSICS

HORACE ROBERT BYERS

discarded

551.576
B991e

THE UNIVERSITY OF CHICAGO PRESS
CHICAGO & LONDON

MERNER - PFEIFFER LIBRARY
TENNESSEE WESLEYAN COLLEGE
ATHENS, TN. 37303

Library of Congress Catalog Card Number: 65-17282

THE UNIVERSITY OF CHICAGO PRESS, CHICAGO & LONDON
The University of Toronto Press, Toronto 5, Canada

© *1965 by The University of Chicago. All rights reserved. Published 1965*
Composed and printed by THE UNIVERSITY OF CHICAGO PRESS
Chicago, Illinois, U.S.A.

APR 7 '70

PREFACE

Since the resurgence of interest in cloud physics stimulated by the demonstration in the late 1940's of the possibility of cloud modification by artificial nucleation, papers on various aspects of cloud physics have appeared in the scientific literature at the rate of several a month. In this flood of new information, much of it representing hitherto undocumented observations, it has been difficult to sort out the facts for summarization in a book. This difficulty confronts the teacher and student as well as the educated scientist who wishes to become acquainted with the field.

The obvious approach is one which takes into account the fact that certain fundamental processes which underly the physics of clouds do not change. Once this framework is established, it is easy to fill in details as they develop. It is with this thought in mind that I have chosen to stress mainly the basic principles in this book.

Two kinds of students are likely to take up the study of cloud physics: physicists and chemists on the one hand, and meteorologists on the other. This book is designed for the latter, who, at least in the United States, far outnumber the former in terms of interest in the subject. Since meteorologists have varied backgrounds in physics and physical chemistry, the book includes a restatement of basic principles where needed.

Two topics that form a part of cloud physics have been omitted in this book. They are cloud electricity and radar analysis of clouds. The first is more appropriately the subject for a book on atmospheric electricity and the second is treated in books on radar meteorology. The dynamics of convection and the analysis of severe local storms derived from cloud systems are mainly left for consideration in special treatises on these topics.

HORACE R. BYERS

v

CONTENTS

1

THERMODYNAMICS
OF MOIST AIR

To study cloud physics one should have a firm grasp of classical thermo-dynamics. A useful way of reviewing general thermodynamics is in its application to the atmosphere, as outlined in this chapter.

In cloud physics a number of concepts of surface chemistry and molecular kinetics are considered which are not normally treated in meteorological thermodynamics. In fact, some of the solutions of thermodynamic problems in conventional meteorology deny the principles on which cloud physics is based. Examples are the two extreme assumptions of reversible and irreversible condensation-abiabatic processes, the one producing cloud and no rain, the other rain and no cloud. These assumptions are accepted on the grounds that between the extremes there is an inconsequential difference in the heat which must be taken up from the condensing particles. But between these two extremes lies the whole study of cloud physics.

In this chapter the thermodynamics of moist air as normally treated in meteorology will be presented. The special physics of clouds will be covered in subsequent chapters.

Equation of State and Virtual Temperature

In all parts of the appreciable atmosphere the gases are well within the range of pressures and temperatures in which their state is represented by the equation for an ideal gas:

$$pv = RT .\tag{1.1}$$

If v is the volume occupied by a mole of the gas, R is a universal constant for all gases (8.3144×10^7 ergs $°K^{-1}$ mole^{-1}). In many meteorological applications the specific volume, $a = 1/\rho = v/m$, where ρ is the density and m is the molecular weight, is used, with the equation of state in the form

$$pa = (R/m)T = R_GT \quad \text{or} \quad p = \rho(R/m)T = \rho R_GT . \quad (1.2)$$

The gas constant R_G refers only to the gas having the molecular weight m.

The equation may also be applied to a mixture of gases, such as exists in the atmosphere, if their proportions are known. In this case a fictitious molecular weight is applied from a weighted mean for all the gases, expressed as

$$\bar{m} = \frac{\Sigma n_i m_i}{\Sigma n_i} = \frac{n_1 m_1}{n} + \frac{n_2 m_2}{n} + \ldots + \frac{n_n m_n}{n} , \quad (1.3)$$

where n_i and m_i are the number of moles and molecular weight, respectively, of the ith component, and n is the total number of moles in the given volume. In the appreciable atmosphere the percentage or fraction by volume of the gases other than water vapor—the so-called dry gases— is constant. Thus, if the dry gases consist of 78 per cent nitrogen, 21 per cent oxygen, and 1 per cent argon by volume, the "molecular weight" of dry air becomes

$$m_d = (0.79 \times 28) + (0.21 \times 32) + (0.01 \times 40) = 28.96 .$$

From Dalton's law of partial pressures the total pressure is given as $P = p_1 + p_2 + \ldots + p_n$, and, since the total mass in a given volume is the sum of the masses of the components, the total density is given by

$$\rho = \rho_1 + \rho_2 + \ldots + \rho_n = (RT)^{-1}(p_1 m_1 + p_2 m_2 + \ldots + p_n m_n) . \quad (1.4)$$

In meteorological practice it is convenient to apply the equation in a form that considers two components, dry air and water vapor. We write

$$\rho = (RT)^{-1}(p_d m_d + p_w m_w) , \quad (1.5)$$

where the subscripts d and w refer to the dry gases and the water vapor, respectively. The practice is to write the equation of state in terms of m_d and the total pressure, which the barometer measures, and to apply a derived correction factor for the water vapor. The total pressure and apparent molecular weight of dry air are placed outside the parentheses of (1.5) to produce the expression

$$\rho = \frac{P m_d}{RT} \left(\frac{p_d}{P} + \frac{p_w}{P} \frac{m_w}{m_d} \right). \quad (1.6)$$

But since $p_d = P - p_w$, we have

$$\rho = \frac{P m_d}{RT} \left[1 + \left(\frac{m_w}{m_d} - 1 \right) \frac{p_w}{P} \right] = \frac{P m_d}{RT} \left(1 - 0.377 \frac{p_w}{P} \right). \quad (1.7)$$

The correction is assigned to the temperature factor by defining a *virtual* temperature

$$T^* = \frac{T}{1 - 0.377 p_w/P} \quad (1.8)$$

and writing the equation of state as

$$\rho = Pm_d/RT^* \quad \text{or} \quad Pa = R_dT^* , \qquad (1.9)$$

etc. In the atmosphere, p_w/P is of the order of 0 to 0.035, so T^* ranges from 0 to 5 degrees higher than T. Differences greater than $3°$ are quite rare, occurring only in tropical air at the bottom of the atmosphere, so the correction usually adds less than 1 per cent to the temperature.

Expressions for the Water-Vapor Content

Under atmospheric conditions water vapor behaves as an ideal gas, so that

$$p_w = \rho_w R_w T . \qquad (1.10)$$

The T needs no subscript because all gases in the mixture at a given location have the same temperature. In addition to the vapor pressure, p_w, and the vapor density, ρ_w, some other expressions for the water-vapor content are useful, for example, specific humidity,

$$q = \frac{\rho_w}{\rho_d + \rho_w} = \frac{R_d}{R_w} \frac{p_w}{p_d + p_w R_d/R_w} = 0.623 \frac{p_w}{P - 0.377 p_w}, \qquad (1.11)$$

mixing ratio

$$w = \frac{\rho_w}{\rho_d} = \frac{R_d}{R_w} \frac{p_w}{p_d} = 0.623 \frac{p_w}{P - p_w}. \qquad (1.12)$$

If both the numerator and denominator of the defining expression for q are divided by ρ_d, it is seen that $q = w/(1 + w)$, and, with p_w/P and therefore w of the order of 10^{-2}, the specific humidity and mixing ratio are numerically essentially the same. It is to be noted that since they are gravimetric measures, they will remain constant during volume changes.

Water vapor is said to be at saturation at a given temperature when it is in equilibrium with a flat surface of pure water at that temperature. The equilibrium is that in which there is no net movement of molecules between the two phases when they are in contact with each other. The saturation values p_{ws}, ρ_{ws}, q_s, and w_s describe the vapor content under these conditions. Supersaturation exists when these values for a given temperature are exceeded, and subsaturation exists when the vapor content is lower than that represented by these values. The saturation ratio, the ratio of the existing to the saturation value at the same temperature, p_w/p_{ws}, etc., is a measure of the degree of subsaturation or supersaturation. The relative humidity is this ratio in percentage form, being defined by the World Meteorological Organization as 100 w/w_s per cent.

From the equation of state it is apparent that at constant temperature $p_w/\rho_w = \text{constant} = p_{ws}/\rho_{ws}$. Since p_{ws} and ρ_{ws} represent the same state (saturation) they must vary together, and each is therefore a function only of temperature. The saturation equilibrium values are higher the higher the temperature. Measurements of the saturation values are made directly and presented in tables, such as the Smithsonian Meteorological Tables. Since undercooling of water below its melting point is common in the atmosphere, the saturation equilibrium condition is determined separately over water and over ice at temperatures below 0° C. It is found that the values over the undercooled liquid are greater than over ice. The great importance to cloud physics of this state of affairs will be emphasized in subsequent pages.

A relatively simple way of measuring the equilibrium vapor pressure at temperatures above freezing is by injecting water into the vacuum end of a Toricellian tube or mercury barometer. The water can be injected in the bottom of the tube through a capillary with care taken not to include air. The water will rise to the top and depress the mercury column by an amount equivalent to the pressure of the vapor (mm Hg) plus the weight of the unevaporated water. The tube is surrounded by a bath for controlling and varying the temperature. Comparisons are made with a tube in the same bath containing only mercury. Readings are made, while noting the temperature of the bath, of the heights of the two mercury columns and of the column of unevaporated water.

Since to each temperature there corresponds a value of vapor pressure representing saturation, one can take the observed vapor pressure and find the temperature for saturation at that vapor pressure. The temperature thus determined is called the temperature of the dew point. One can define it as the temperature to which the air must be cooled at constant vapor pressure to produce saturation; or the cooling may be specified as taking place at constant total pressure and mixing ratio. Given the dew-point temperature, one can obtain the vapor pressure from the saturation tables and, with the temperature, all of the other water-vapor quantities. A variety of combinations of two vapor quantities or of a humidity and a temperature will produce the others.

It is convenient to have a formula for the saturation vapor pressure as a function of temperature which can be inserted in various physical equations to avoid cumbersome numerical computations from the tables. The thermodynamically derived Clapeyron-Clausius equation, to be discussed on subsequent pages, is quite suitable for most purposes. In the Smithsonian Meteorological Tables the reader will note that the World Meteorological Organization has adopted a less idealized but quite cumbersome version of this equation. Where great accuracy or absolute values are wanted, the tables should be used, but for relative values and changes through small ranges the idealized equation is entirely adequate.

A situation often exists where vapor and liquid are in contact but not in equilibrium. They may not even be at the same temperature. The force at the liquid surface driving molecules from it into the surrounding space depends on the temperature of the liquid surface and, for unit area, is exactly equal to the saturation vapor pressure at that temperature. It is desirable to avoid confusion by speaking of the force per unit area exerted by the liquid as the *vapor tension* of the surface and using the term vapor pressure only for the partial pressure of the gas. The vapor is then said to be in equilibrium with the liquid or ice surface when its vapor pressure is equal to the vapor tension of the surface.

We shall return later to a consideration of the thermodynamic properties of water.

Some General Thermodynamic Relations

The laws of thermodynamics provide a statement of the thermodynamic balance as

$$T \, dS = dE - dW \, , \tag{1.13}$$

where S, E, and W are the entropy, the internal energy, and the work done on the system, respectively. For a gas, the balance per mole is

$$T \, dS = dE + p \, dv = C_v dT + p \, dv \, , \tag{1.14}$$

where C_v is the heat capacity per mole. Per gram, the balance is

$$T \, ds = c_v dT + p \, da \, , \tag{1.15}$$

where c_v is the specific heat at constant volume $(= m C_v)$. We can write the equation of state in differential form, that is, $p \, dv + v \, dp = R \, dT$. Substituting for $p \, dv$ in (1.14) and remembering that $C_v + R = C_p$, we obtain

$$T \, dS = C_p dT - v \, dp \, . \tag{1.16}$$

Or, we may write

$$T \, ds = c_p dT - a \, dp \, , \tag{1.17}$$

where c_p is the specific heat at constant pressure.

These last two forms of the thermodynamic balance equation are useful in studying gases in the open atmosphere where we are not dealing with fixed or identifiable volumes.[1]

In many situations in the atmosphere, heat is gained or lost at constant pressure, such as through absorption of radiation, but of greater interest for cloud studies are those processes involving pressure changes. The decreasing pressure with height results in marked thermal effects on air parcels that have a vertical component of motion.

If no heat is added or removed, the process is isentropic or adiabatic, and

$$C_p dT - v \, dp = 0 = c_p dT - a \, dp \, , \tag{1.18}$$

[1] Henceforth we will use these and related expressions for atmospheric conditions, writing the total pressure as p, without using virtual temperature except in those cases where it might be significant. In other words, the atmosphere will usually be considered as made up of a mixture of gases in constant ratio to each other, having the total pressure p and the "molecular weight" m of dry air.

$$\frac{dT}{dp} = \frac{v}{C_p} = \frac{RT}{C_p p} = \frac{RT}{m\,c_p p} = \frac{a}{C_p}. \tag{1.19}$$

In air moving adiabatically across the pressure surfaces, therefore vertically, at a speed $\omega = dp/dt$, positive downward, toward increasing pressure,

$$\frac{dT}{dt} = \frac{dT}{dp}\frac{dp}{dt} = \omega\,\frac{RT}{m\,c_p p} = \omega\,\frac{a}{c_p}. \tag{1.20}$$

Except in the case of very strong buoyancy accelerations, the atmosphere may be assumed to be in hydrostatic equilibrium, with the pressure related to height by the hydrostatic equation

$$dp = -\,\rho g dz = -\,\rho d\Phi\,, \tag{1.21}$$

where g is the acceleration of gravity, z is height, and Φ is the geopotential ($d\Phi = gdz$, or the work done in lifting one gram through the height dz). If the static air density is the same as that of the vertically moving air, substituting the value dp in (1.19) gives

$$\frac{dT}{dz} = -\frac{g}{c_p}, \qquad \frac{dT}{d\Phi} = -\frac{1}{c_p}. \tag{1.22}$$

Parcels of air undergoing adiabatic temperature changes in vertical motions often have temperatures different from those of the static environment. This difference is to be expected because the observed temperature lapse rates are not in adiabatic equilibrium. One should then write

$$\frac{dT}{dz} = -\frac{g}{c_p}\frac{\rho}{\rho'} = -\frac{g}{c_p}\frac{T'}{T}, \qquad \frac{dT}{d\Phi} = -\frac{1}{c_p}\frac{T'}{T}, \tag{1.23}$$

where the primed quantities refer to the moving parcel.[2]

It is seen that the T'/T factor does not enter when the motion is described in terms of pressure change, because the hydrostatic pressure itself adjusts to changes in static density, and the pressure inside and outside the parcel must be the same.

To obtain the temperature after a finite adiabatic displacement from p_0, T_0 to p, T, one can take equation (1.17) for the case of $ds = 0$ in the form

$$c_p\frac{dT}{T} = \frac{a}{T}\,dp = \frac{R}{m}\frac{dp}{p} \tag{1.24}$$

---

[2] The use of virtual temperatures in these expressions is an unnecessary nicety, since in subsaturated conditions both inside and outside the parcel the p_w/P ratios are about the same. Another refinement has to do with the specific heat, since c_p for the dry gases is 0.24 and for water vapor 0.441 cal $g^{-1}\,^\circ K^{-1}$ (to be multiplied by the mechanical equivalent of heat to obtain the work units normally used). Instead of c_p the denominator should contain $wc_p' + c_p$, where c_p' is the specific heat of water vapor. Again, with w of 10^{-3} to 10^{-2}, the effect of the water vapor can be neglected.

and integrate it; the resulting

$$\ln \frac{T}{T_0} = \frac{R}{m\,c_p} \ln \frac{p}{p_0} \tag{1.25}$$

can be written as

$$T = T_0(p/p_0)^k \,, \tag{1.26}$$

where $k \equiv R/mc_p = 0.286$.

The potential temperature is defined as the temperature reduced abiabatically to a pressure of one bar, or

$$\theta = T(1/p)^k \tag{1.27}$$

for p expressed in bars, but any pressure units may be used.

By definition, the potential temperature is constant in an adiabatic process. Since the common situation in the atmosphere is one in which the temperature is observed to decrease with decreasing pressure (increasing height) at a rate different from that of the adiabatic process, usually at a lower rate, it is useful to form the relationship between the vertical distributions of temperature and potential temperature. They can be derived by expressing (1.27) in the logarithmic differential form

$$\frac{d\theta}{\theta} = \frac{dT}{T} - \frac{R}{m\,c_p}\frac{dp}{p} \,, \tag{1.28}$$

from which it follows that

$$\frac{d\theta}{dp} = \frac{\theta}{T}\left(\frac{dT}{dp} - \frac{RT}{m\,p\,c_p}\right) = \frac{\theta}{T}(\gamma_p - \Gamma_p), \tag{1.29}$$

where Γ_p is seen as the adiabatic rate derived in equation (1.19). Substitution for dp from the hydrostatic equation results in

$$\frac{d\theta}{dz} = \frac{\theta}{T}\left(\frac{dT}{dz} + \frac{g}{c_p}\right) = \frac{\theta}{T}(\gamma - \Gamma), \tag{1.30}$$

where $\gamma = -dT/dz$ and Γ is the adiabatic rate $-g/c_p$ of equation (1.22). It is apparent that θ increases with height when the temperature lapse rate is less than the adiabatic and decreases with height in the less common superadiabatic lapse rate.

Adiabatic Processes with Vapor Condensation

In cloud physics we are concerned with vertical motions in clouds where the adiabatic process is modified by the exchange of heat through condensation and evaporation of water. In condensation, the amount of water vapor in a gram of dry air will change by $-dw_s$. The heat produced in condensation of this amount of vapor is

$$\delta Q = - d(Lw_s) \,, \tag{1.31}$$

where L is the heat released by condensation of one gram from the vapor. The L decreases slowly with temperature while w_s increases rapidly with temperature

7

and decreases slowly with increasing pressure.[3] In the ascent of the $(1 + w_s)$-gram parcel there will be a certain mass of liquid water, χ, carried along from the warmer levels below. It will give off the sensible heat $-s\chi \, dT$, where s is its specific heat. Meanwhile the $1 + w_s$ grams of air is expanding in a manner that is adiabatic except for these two forms of heat, so the balance is

$$- d(Lw_s) - s\chi \, dT = (1 + w_s)\left(c_p dT - \frac{RT}{m}\frac{dp}{p}\right). \qquad (1.32)$$

An examination of the various terms shows that the w_s on the right adds less than 1 per cent to the difference which follows. We can take c_p and m as representing only the dry gases and use the partial pressure p_a; thus, with w_s neglected, the terms on the right would represent dry air. The liquid-water term, that is, the second term on the left, might have an extreme high value of 5 per cent of $c_p dT$.[4] In exact calculations the liquid-water effect should not be overlooked if there are large amounts of liquid water $(\chi \times 10^3 = 5 \text{ g kg}^{-1} \sim 5 \text{ g m}^{-3})$. We may then find suitable accuracy in the following equation:

$$(s\chi + c_p) \, dT - \frac{RT}{m}\frac{dp_a}{p_a} + d(Lw_s) = 0, \qquad (1.33)$$

and for low liquid-water content

$$c_p dT - \frac{RT}{m}\frac{dp_a}{p_a} + d(Lw_s) = 0. \qquad (1.34)$$

Thus, it is seen that the process is in most cases adequately described by adding a term $d(Lw_s)$ to the dry-adiabatic process.

Equation (1.34) essentially assumes that no liquid water is present in the parcel or, as one might say, that all of the water is precipitated out as fast as it is condensed. Physically, this is a preposterous assumption, but in the thermodynamic calculation it does not distort the result. The process under this assumption is called pseudoadiabatic. The exact expression for it is given by equation (1.32), with $\chi = 0$ and with p_a used. Only if liquid water is present can the process become reversible. The reversible process is that brought about by compression in descending air, and then w_s must increase, which it can do

[3] In the atmospheric range w_s always decreases with ascent and increases with descent. Since it represents water-vapor saturation, the required w_s will not occur in descent unless there is a supply of water droplets or ice particles with which saturation equilibrium can be maintained. Note that $w_s = 0.623 \, p_{ws}/p_a = f(T)/p_a$, and that it is much more sensitive to changes in temperature than to changes in atmospheric pressure.

[4] The maximum values of χ might be expected in a parcel in which all of the condensed water is conserved and carried along. We would then have $\chi = w_{so} - w_s$, where w_{so} is the saturation mixing ratio at the condensation level, presumably the cloud base. In such a parcel, χ would increase steadily with height.

only by the evaporation of the liquid. The dw_s is positive and L is the same as before but of opposite sign (heat taken up from the air), so the last term in (1.33) is negative. In the pseudoadiabatic case, the descent is along the dry adiabatic because there is no liquid water to evaporate and maintain saturation in the parcel.

In the case of direct deposition from the vapor to the solid, the so-called "snow stage," equation (1.33) becomes

$$(s'\chi' + c_p) \, dT - \frac{RT}{m} \frac{d p_d}{p_d} + d (L_s w_s) = 0 , \qquad (1.35)$$

where $s'\chi'$ is a weighted mean specific heat and water content combining both liquid and ice, and L_s is the heat of sublimation.

A "hail stage" in which the water contained in an ascending parcel passes from liquid to ice is also defined. The thermodynamics of this stage rests on assumptions concerning heat exchange between the freezing water and the cloud air which may bear little resemblance to reality. A discussion of hail-forming and hail-growth processes will be left for a later section.

In thermodynamic diagrams for meteorological uses the pseudoadiabatic lines are constructed on the basis of the vapor-liquid transition only, even at temperatures below the melting point of water (0° C). The prevalence of under-cooled water in clouds down to temperatures of −20° C or less is taken as justification for this usage. At very low temperatures, $(\partial w_s / \partial p)_T$ and $(\partial w_s / \partial T)_p$ are quite small and L differs from L_s by less than 10 per cent.[5]

One can assign values to the pseudoadiabatic lines which have a thermodynamic meaning. Two potential temperatures have been defined to characterize a pseudoabiabatic line. They are (1) the potential temperature which is approached as w approaches zero in a long-continued pseudoadiabatic expansion, called the *equivalent-potential temperature*, and (2) the temperature at which the pseudoadiabat intersects the 1000-mb pressure line, called the *wet-bulb potential temperature*. The latter derives its name from the fact that it involves the same type of heat exchange as that occurring around a wet-bulb thermometer. In the following paragraphs the second of these two quantities will be treated first.

Wet-Bulb Temperature

The usual way of measuring humidity at weather stations is by means of a psychrometer, which is simply an ordinary thermometer and a wet-bulb thermometer mounted together. The wet bulb differs from the dry in having a fitted piece of untreated cotton cloth, such as muslin, covering it, so that when

[5] For example, at −35° C and 500 mb the pseudoadiabatic rate in the vapor-liquid stage is 0.900 of the dry adiabatic, and in the vapor-ice stage it is 0.909 of the dry adiabatic; at −20° C and 700 mb the values are 0.814 and 0.808, respectively. The two are identical at about −25° C from about 850 mb upward.

wetted with water it remains wet long enough to permit the taking of an observation. Evaporation of water from the wet bulb cools it to an equilibrium temperature which occurs when the heat removed from the bulb by evaporation is balanced by the sensible heat being taken up by the bulb from the air. The wet-bulb temperature thus depends on the temperature and vapor pressure in the air, and the readings of the two thermometers comprising the psychrometer can uniquely determine the humidity. The relationship has been determined exactly through empirical means by such workers as Ferrel (1886), but the direct thermodynamic definition has important meteorological uses. The result is a thermodynamic wet-bulb temperature which does not differ materially from that measured by the wet-bulb thermometer. It will now be treated in a way that is applicable to a water drop as well as a wet bulb. Generally adopted derivations and definitions are due to Normand (1921), Bleeker (1939), and Petterssen (1956).

The removal of water vapor from the wet bulb by evaporation can be characterized by a coefficient of diffusivity of water vapor in air, D (cm^2 sec^{-1}). Since the psychrometer is ventilated, it is an eddy type of diffusivity. The flux of heat from the ambient air to the bulb is characterized by a coefficient of thermal diffusivity, κ (also cm^2 sec^{-1}). There is a small exchange of heat through radiation, as shown by Dropkin (1939), which can be taken into account by adjusting the value of the coefficient.

The flux of vapor from a unit area of the bulb is given by

$$\frac{dM}{dt} = D(\rho_{w0} - \rho_{w\infty}),\qquad (1.36)$$

where the subscript 0 refers to conditions at the bulb and ∞ refers to the ambient air. The flux of latent heat per unit area is given by

$$\frac{dQ_L}{dt} = L\frac{dM}{dt},\qquad (1.37)$$

where L is the latent heat of vaporization. At the wet-bulb temperature this heat loss is balanced by the gain of sensible heat from the ambient air which, expressed as a flux per unit area, is given by

$$\frac{dQ_s}{dt} = \rho\kappa c_p(T - T_w) = \frac{dQ_L}{dt} = L\frac{dM}{dt} = LD(\rho_{w0} - \rho_{w\infty}),\qquad (1.38)$$

$$\frac{T - T_w}{\rho_{w0} - \rho_{w\infty}} = \frac{LD}{\rho\kappa c_p},\qquad (1.39)$$

where T_w is the wet-bulb temperature, ρ is the air density, and c_p the specific heat of the air. At ventilation rates of 2.5 m sec^{-1} or greater,[6] D and κ apparently

[6] The WMO recommends a ventilation rate of the psychrometer of 4 to 10 m sec^{-1}. Note also that in this case D and κ are eddy diffusivities.

are about the same and cancel each other. Since $\rho_w/\rho = q$, we may write

$$T - T_w = \frac{L}{c_p}(q_0 - q_\infty) \simeq \frac{L}{c_p}(w_0 - w_\infty),\tag{1.40}$$

and the wet-bulb temperature is given by

$$T_w = T - \frac{L}{c_p}(w_0 - w_\infty).\tag{1.41}$$

Since w_0 is the mixing ratio for saturation at the temperature T_w at the given dry-air pressure, it is readily obtained. The actual mixing ratio of the air is found from a psychrometric observation by solving for w_∞. In practice, tables are used for obtaining the various humidity quantities from psychrometric readings.[7]

As stated previously, the same thermodynamics is applicable to water drops suspended or falling in the air. If there are a large number of water drops or ice particles evaporating or subliming, the air will be cooled while its humidity is increased. The air exchanges sensible heat for latent heat. This exchange stops when the air and the particles are at the same temperature and the particles are no longer evaporating or subliming, which is possible only when saturation is reached. This limiting saturation equilibrium occurs at the wet-bulb temperature. The mixing ratio in the air is the same as the equilibrium mixing ratio over the drops, that is, $w_\infty = w_0$ and $T = T_w$.

From this discussion it is clear that the wet-bulb temperature must be higher than the dew-point temperature because the mixing ratio has been increased to produce the former and is specified as constant in the definition of the latter. The higher the mixing ratio the higher will be the temperature at saturation.

The thermodynamic wet-bulb temperature produced by a condensation-adiabatic or pseudoadiabatic process is the most useful one to consider because it provides a means of assigning meaningful values to the pseudoadiabatic lines. The process may be described as follows.[8] Lift a sample parcel dry adiabatically to its condensation point, determined by its potential temperature and mixing ratio. Then add enough liquid water in the form of cloud drops to keep the air parcel saturated by evaporation of these drops during descent to the original level. If there is more than the required amount of water the result will not be affected. With the water continually evaporating, the parcel will descend along a reversible saturation-adiabatic line, which is, numerically, essentially the same as a pseudoadiabatic line. Where this line intersects the line of the original pres-

[7] For a discussion of humidity standards and standard instruments, the reader should refer to the text in the Smithsonian Meteorological Tables (1951) and the references given therein.

[8] In reviewing processes described on these pages the student should have at hand one of the thermodynamic diagrams described at the end of the section in this chapter entitled *Work Diagram*.

11

sure, the wet-bulb temperature is achieved. This temperature has been called the pseudo–wet-bulb temperature by Petterssen (1956) and has been treated, although not in each case with the same nomenclature, by Normand (1931), Bleeker (1939), and Rao (1945, 1957, 1960). It is usually 0.1° C or so lower than the measured T_w.

The wet-bulb potential temperature, sometimes written with the prefix *pseudo*, is defined by the same process carried to the 1000-mb pressure. It is seen that each saturation-adiabatic or pseudoadiabatic line corresponds to one wet-bulb potential temperature as indicated by its intersection with the 1000-mb line. Since there can be only one pseudoadiabat going through a given condensation point and since the condensation point is uniquely determined by the potential temperature and mixing ratio of the parcel, the wet-bulb potential temperature is uniquely determined in the same way. The adiabatic condensation point is sometimes called the *characteristic point*, because from it all manner of thermodynamic processes in the atmosphere can be determined.

The wet-bulb temperature defined by (1.41) is achieved at constant dry-air pressure, while the pseudo–wet-bulb temperature just described results from a type of adiabatic pressure change. The quantitative difference between the two may be noted by writing (1.41) in differential form as

$$c_p dT = - L \, dw \tag{1.42}$$

and comparing it with the pseudoadiabatic equation in a form equivalent to (1.34):

$$c_p dT = \frac{RT}{m} \frac{dp_d}{p_d} - L \, dw \tag{1.43}$$

with L regarded as constant. The difference is in the first term on the right which, when integrated up to the characteristic point and back, represents the small amount of work expended in the process. Rao (1960) has computed and tabulated the difference in the resulting temperatures and has found that the pseudo–wet-bulb temperature is lower than that achieved in the constant-pressure process by less than 0.5° C, except in very dry, warm air where the difference may reach 1° C.

Equation (1.43) divided by T as in (1.34) may be integrated downward from the characteristic point if T as a function of w_s, p_d is substituted. The process is easily followed on a suitable thermodynamic diagram to obtain the result graphically.

Equivalent-Potential Temperature

If a pseudoadiabatic expansion is continued to a low pressure, the water vapor condenses out until the $L \, dw$ term disappears and (1.34) becomes the dry-adiabatic equation for dry air. Since the dry-adiabatic process defines a potential temperature, the pseudoadiabat finally becomes a potential-temperature line. This line represents the equivalent-potential temperature of all air samples having characteristic points on the given pseudoadiabat.

Rossby (1932) derived the expression for the equivalent-potential temperature in terms of the potential temperature of the dry air and the mixing ratio. The dry-air potential temperature θ_d is used because (1.34) contains the partial pressures of the dry air and the potential temperature inserted in such an equation would have to be related to these pressures. Actually it differs from the total potential temperature by 1° C or less.

From (1.25) the potential-temperature equation is written as

$$\ln T = \frac{R}{m_d c_p} \ln p_d - \frac{R}{m_d c_p} \ln 1000 + \ln \theta_d. \tag{1.44}$$

This expression for $\ln T$ is substituted in (1.43) to eliminate the $\ln p_d$ term resulting in

$$\ln \theta_d - \frac{R}{m_d c_p} \ln 1000 + \frac{Lw}{c_p T} = \text{const}. \tag{1.45}$$

As $w \to 0$, $\theta_d \to \theta_E$, the equivalent-potential temperature, so that

$$\ln \theta_E - \frac{R}{m_d c_p} \ln 1000 = \text{const}, \tag{1.46}$$

and, consequently,

$$\ln \theta_E = \ln \theta_d + (Lw/c_p T), \tag{1.47}$$

$$\theta_E = \theta_d \exp (Lw/c_p T). \tag{1.48}$$

The θ_d, w, and T may be the values taken anywhere in the pseudoadiabatic process, which is taking place along the line $\theta_E = \text{const}$. For the unsaturated case, θ_d and w are constant in an expansion up to the condensation point, and the start of the integrated equation is at the characteristic point. One then uses the temperature of the characteristic point, T_c, and writes the expression in the more general form,

$$\theta_E = \theta_d \exp (Lw/c_p T_c). \tag{1.49}$$

The vertical distribution of the equivalent and wet-bulb potential temperatures provides an indication of the possible development of instability. Frequently the unsaturated lower and middle troposphere is observed to have decreasing values of these quantities with height. If the layers should become saturated, they would provide instability for a parcel moving along a pseudoadiabat ($\Gamma_s < \gamma$). Deep layers can become saturated through large-scale ascending motions or through evaporation of precipitation falling from an upper cloud sheet.

Work Diagram

From elementary thermodynamics it is shown that the work done in carrying a gas through a closed cycle of changes of state is given by the integral over the closed path

$$W = \int_c p \, da. \tag{1.50}$$

The diagram with coordinates of p and v or p and a is a work or energy diagram, since the integral in these coordinates around a closed path is equal to the area enclosed by the path and work is given by an area on the diagram.

In meteorology, as indicated before, it is easier to operate with pressure and temperature than with pressure and volume. By substitution from the equation of state in differential form, it is seen that

$$\int_c p \, da = \int_c \frac{R}{m} \, dT - \int_c a \, dp = \int_c \frac{R}{m} \, dT - \int_c \frac{RT}{m} \frac{dp}{p}. \tag{1.51}$$

But the first integral on the right is zero because, with temperature as the only variable, the path goes out and back along the same line and no area is enclosed. Then

$$W = \int_c p \, da = -\frac{R}{m} \int_c T \, d(\ln p). \tag{1.52}$$

A diagram with the coordinates T and $\ln p$, with p increasing toward the T-axis, is used in meteorology. It is an exact transformation of the p,a diagram with a constant transformation factor R/m.[9]

An approximation of the T, $\ln p$ diagram developed by Stüve (1927) is one with T and p^k, where k is the constant in the adiabatic equation. The lines of potential temperature are straight on this diagram, an advantage which for many purposes outweighs the disadvantage of a slight area distortion.[10]

A complete diagram has lines of temperature, pressure, potential temperature, saturation mixing ratio, and pseudoadiabats. Exercises with these charts usually are included in elementary courses in meteorology.

Buoyancy and Work

Parcels move upward or downward depending on their relative buoyancy. The buoyant acceleration in a gravitational field is given by the familiar expression of elementary physics

$$M'a = g(M - M'),$$

[9] A number of other equal-area transformations of the p,v diagram are used; for example, the tephigram (Shaw, 1926), the aerogram (Refsdal, 1935), and the skew T, $\ln p$ diagram (Herlofson, 1947).

[10] The potential-temperature lines on a T, $\ln p$ diagram are so nearly straight that one has to sight along the plane of the paper to notice their curvature.

where a is the upward acceleration of the parcel of mass M' and M is the mass of the air it displaces. Since the volumes of the parcel and the displaced air are the same,

$$a = g \frac{\rho - \rho'}{\rho'} = g \frac{T' - T}{T} = g \left(\frac{T'}{T} - 1 \right). \qquad (1.53)$$

As an isolated parcel moves, it displaces "environment" or "ambient" air having the temperature T as measured by a sounding. At any level the acceleration is proportional to the ratio of the parcel temperature to that of the environment at that level.

If a parcel starts its ascent or descent with the same temperature, T_0, as its environment and changes temperature at either a dry or saturation rate, then at height Δz above or below the starting point the acceleration is

$$a = \frac{(\Gamma - \gamma) \Delta z}{T_0 + \gamma \Delta z}, \qquad (1.54)$$

where Γ is the dry or saturation adiabatic rate, as the case may be, and γ is the environment lapse rate. If the lapse rate is less than the adiabatic (dry or saturated), the acceleration will be opposite to the sign of Δz, thus restoring the parcel to z_0, a stable condition. If the lapse rate is greater than adiabatic, the acceleration is in the direction of the initial displacement and hence unstable. In the latter case, the accelerations increase with the displacement Δz, whether upward or downward.

The work done in the finite displacement of a parcel of unit mass in buoyant acceleration is given by

$$W = \int_{z_0}^{z} a \, dz = \int_{z_0}^{z} g \frac{\rho - \rho'}{\rho'} \, dz = \int_{z_0}^{z} g \frac{\rho}{\rho'} \, dz - \int_{z_0}^{z} g \, dz, \qquad (1.55)$$

and, substituting from the hydrostatic equation and realizing that $p = p'$, $dp = dp'$, we have

$$W = - \int_{p_0}^{p} \frac{dp}{\rho'} + \int_{p_0}^{p} \frac{dp}{\rho} = - \frac{R}{m} \int_{p_0}^{p} T' \frac{dp}{p} + \frac{R}{m} \int_{p_0}^{p} T \frac{dp}{p}$$

$$= \frac{R}{m} \int_{p}^{p_0} (T' - T) \, d(\ln p). \qquad (1.56)$$

Thus it is seen that this work is given on the T, $\ln p$ diagram by the area between the temperature curve of the moving parcel and the environment lapse-rate curve. The unstable case ($T' > T$ in ascent, $T' < T$ in descent) is of the most practical interest, because in the stable case large displacements are impossible. The unstable case applies to both saturated and dry conditions, but instability in moist air is the important situation, except in desert regions. In the usual saturated unstable distribution the area between the curves is closed

15

by intersections at the top and the bottom. The basic test for instability is performed by hypothetically displacing a parcel sampled from the air and measuring the positive and negative areas. Instability that is realized only after the parcel passes through a small low-level negative area and, following a pseudoadiabatic path, forms a positive area, is called conditional instability. This type of instability is characteristic of cloud-forming heat convection in the atmosphere.

In determining the buoyancy it is necessary to use the virtual temperature because in the relevant cases the ascending air reaches saturation while the environment is relatively dry. Virtual temperature differences of the order of $1°$ C may correspond to strong updrafts; a simple computation shows that the unresisted acceleration produced by a $1°$ temperature difference would result in a 10 m sec^{-1} updraft in less than 2500 m of ascent.

In the real cloud case it is necessary to take into consideration the liquid water, not so much in the sense of its thermal effects as expressed in equation (1.33) which would tend toward a higher temperature, but in terms of the actual weight of the water as it affects the buoyancy. Saunders (1957) has introduced a "cloud virtual temperature" which is based on the density obtained by including the liquid as well as the vapor. The liquid changes the density by an amount $\Delta\rho' = \rho' \cdot \chi$ and the virtual temperature by an amount $-\Delta T^{*'} = T^{*'} \cdot \chi$. Although the liquid in the reversible saturation-adiabatic process, equation (1.33), makes the temperature in ascent higher than in the pseudo-adiabatic case, the cloud virtual temperature as defined by Saunders is less than that produced by pseudoadiabatic ascent. For example, $\Delta T^{*'} \approx -1°$ C for an ascent of 200 mb in summer tropical air. As will be noted in the next section, however, the liquid water content very rarely is as high as the parcel-ascent value because of mixing with environment air.

The resistance to buoyancy motions cannot be treated in a straightforward manner. The velocity profile across a vertical current is usually sharp, that is, it has pronounced shear, and shear in a fluid or gas always results in a transfer of momentum by turbulent eddies. The effect has the character of frictional resistance. The form drag on the cloud envelope, as on a bubble or thrusting spearhead, also may enter into the problem.

Vertical motions must always develop some form of air circulation as part of the general convection pattern. Continuity of mass requires that in order to prevent accumulation or attenuation of mass the total divergence of the velocity must be zero, or $\nabla \cdot \mathbf{C} = 0$. It is convenient to separate the divergence into two components, the horizontal or two-dimensional divergence, $\nabla_2 \cdot \mathbf{C}$, and the vertical divergence, $\nabla_z \cdot \mathbf{C}$, and to write $\nabla_2 \cdot \mathbf{C} + \nabla_z \cdot \mathbf{C} = 0$. In this way one can note that horizontal convergence (negative divergence) is associated with vertical divergence. At the ground, where the vertical velocity is zero, horizontal convergence means upward acceleration, $-\nabla_2 \cdot C = \nabla_z \cdot C = \partial C_z/\partial z$ where the positive z-direction is upward.

Entrainment

As anyone acquainted with the ways of nature might guess, parcels of air cannot remain separate from their environment: mixing of ambient air must occur. In view of this reality, the parcel method is merely a starting point as a definitive test of instability representing a limiting case. Convection currents in laboratory-fluid models invariably show jetlike upward motion from small heat sources, with entrainment of ambient fluid into the current, as is characteristic of jets. Whether the convection is considered as being in the form of cylinders or fountains of fairly continuous upward flow or in the form of pulsating bursts or "bubbles" makes no difference with respect to the basic concepts of entrainment.

In the type of convection which gives rise to cumulus and cumulonimbus clouds, a certain amount of mixing between the cloud air and the environment air occurs. It has been shown that in order not to be decelerated in its upward growth a cloud core must be warmer than its surroundings. The mixing of colder environment air into the cloud has the effect of reducing the temperature of the rising air. Since the outside air is dry, its mixing with the cloud air would produce subsaturation, but saturation is maintained by evaporation of some of the liquid water already present in the cloud. This evaporation cools the mixture further. In this way the buoyancy is reduced. The greater the proportion of outside air in the mixture and the lower its initial water-vapor content the greater will be the cooling.

Austin (1951) showed that for environment lapse rates less than the dry adiabatic, the cloud temperature will become lower than that of the environment before the liquid water is exhausted, so that in the usual case dissipation of a cloud results from loss of buoyancy rather than from drying up in the mixing.[11] Negative buoyancy initiates downward motion which causes the liquid water to be used up, because w_s increases rapidly downward along a pseudoadiabat.

Stommel (1947) was the first to point out the significance of entrainment in convection. He suggested that the effects could be computed graphically in steps. The steps are best stated as follows:

(1) Lift a 1-kg sample parcel of air from its characteristic point through a pseudoadiabatic expansion some chosen small distance to a new T',p.

(2) Mix this with M kg of environment air having the temperature T and the specific humidity q to make a mixture of $(1 + M)$ kg. Leave the evaporation effects for the next step. The mixture would now have the temperature $T_m =$

[11] Austin's mathematical proof can be verified graphically if one realizes that the mixture will be cooled by the evaporation to its pseudo–wet-bulb temperature.

17

$(T' + MT)/(1 + M)$ and the specific humidity $q_m = (q' + Mq)/(1 + M)$, where the primed quantities refer to the parcel before mixing. It should be noted that q_m is a subsaturated value.

(3) Evaporate enough of the cloud's liquid water to keep the mixture at saturation. The resulting temperature will be the wet-bulb temperature of the mixture, or

$$T'' = T_m - \frac{L}{c_p}(q'' - q_m), \qquad (1.57)$$

where T'' and q'' are the actual temperature and specific humidity, respectively, resulting in the cloud.[12]

From this T'', p, another parcel-ascent step is taken followed by mixing with the environment at the new level, and so on, until the ascending air irrecoverably loses its buoyancy.

The amount of liquid water added to an isolated parcel going from point 1 to point 2 is given as $\chi_2 - \chi_1 = q_{s1} - q_{s2}$, where χ and q_s are the mass ratio of liquid water to air and the saturation specific humidity, respectively. The values of q_s or, strictly speaking, of w_s can be read directly from the thermodynamic diagrams. The amount of liquid water lost by evaporation in the mixing is $q'' - q_m$.

Ample observational evidence has been accumulated from airplane flights of the Chicago Cloud Physics group and others to show that entrainment is an important factor in convective clouds. Updrafts are found to be warmer than the environment by only 1° C or less in most growing cumulus clouds where the parcel method would produce differences of 2° or 3° C. Normally the liquid water contents are markedly lower than those to be expected from undiluted parcel ascent. Occasional situations are found in which the liquid water content exceeds the parcel-ascent value, suggesting some form of raindrop water storage. Temperatures in updraft cores of large growing cumulus-cloud masses over North America in summer may, under favorable circumstances, approximate parcel-ascent values.

The structure and dynamics of clouds will be taken up again in chapter 7.

[12] Note that the mixing ratio, w, and the specific humidity, q, are considered equivalent, as justified earlier.

GENERAL WORKS

BYERS, H. R. (1959), *General Meteorology* (3d ed.; New York: McGraw-Hill Book Co.).

GODSKE, C. L., BERGERON, T., BERKNES, J., BUNDGAARD, R. C. (1957), *Dynamic Meteorology* (Boston and Washington: American Meteorological Society and Carnegie Institution).

HESS, S. (1959), *Introduction to Theoretical Meteorology* (New York: Holt, Rinehart and Winston).

HOLMBOE, J., FORSYTHE, G. E., GUSTIN, W. S. (1945), *Dynamic Meteorology* (New York: John Wiley and Sons).

REFERENCES

AUSTIN, J. M. (1951), *American Meteorological Society Compendium of Meteorology*, ed. J. F. MALONE (Boston: American Meteorological Society), p. 694.

BLEEKER, W. (1939), *Quart. J. Roy. Met. Soc.*, **65**, 547.

DROPKIN, D. (1939), Cornell Univ. Eng. Exper. Station Bull. No. 26.

FERREL, W. (1886), *U.S.A. Ann. Rep. Chief Signal Officer*, Appendix 24, p. 233.

HERLOFSON, N. (1947), *Met. Ann.* (Oslo), **2**, No. 10.

NORMAND, C. W. B. (1921), *Mem. Indian Met. Dept.*, **23**, Pt. 1, 1.

———. (1931), *Nature*, **128**, 583.

PETTERSSEN, S. (1956), *Weather Analysis and Forecasting* (2d ed.; New York: McGraw-Hill Book Co.), **2**, 42.

RAO, K. N. (1945), *Proc. Nat'l. Inst. Sci. India*, **11**, No. 2, p. 157.

———. (1957), *J. Met. Soc. Japan, 75th Anniv. Vol.*, p. 87.

———. (1960), *Beitr. Phys. Atmos.*, **32**, 202.

REFSDAL, A. (1935), *Met. Zeitschr.*, **52**, 1.

ROSSBY, C. G. (1932), *Mass. Inst. Tech. Papers in Met.*, **1**, No. 3.

SAUNDERS, P. M. (1957), *Quart. J. Roy. Met. Soc.*, **83**, 342.

SHAW, N. (1926), *Manual of Meteorology* (Cambridge: Cambridge University Press), **1**, 266. See also ed. of 1942, **3**, 269–300.

SMITHSONIAN INSTITUTION (1951), *Smithsonian Meteorological Tables* (6th ed.; Washington, D.C.: Smithsonian Miscellaneous Collection, Vol. **114**).

STOMMEL, H. (1947), *J. Met.*, **4**, 91.

STÜVE, G. (1927), *Beitr. Phys. freien Atmos.*, **13**, 218.

19

2

EQUILIBRIUM AND CHANGE OF PHASE

In considering phase equilibria and phase changes it will be useful to review first some thermodynamic principles employed in physical chemistry.

The Thermodynamic Potentials

Three potentials, largely due to Gibbs (1875), are in use. They are

$$H = E + pv ,\tag{2.1}$$

$$W = E - TS ,\tag{2.2}$$

$$G = E - TS + pv .\tag{2.3}$$

The first is called the heat content or *enthalpy*, the second is the work potential or Helmoltz free energy, and the third is the Gibbs free energy, sometimes called the Gibbs potential, the available energy, or, in this book, simply the free energy. In differential form they are

$$dH = dE + p\ dv + v\ dp = TdS + v\ dp ,\tag{2.4}$$

$$dW = dE - TdS - SdT = - SdT - p\ dv ,\tag{2.5}$$

$$dG = dE - TdS - SdT - p\ dv + v\ dp = v\ dp - SdT .\tag{2.6}$$

In each case the value of dE is substituted from equation (1.14). They are all exact differentials, which means that, considering the three in the order given above,

$$(\partial T/\partial p)_S = (\partial v/\partial S)_p ,\tag{2.7}$$

$$(\partial S/\partial v)_T = (\partial p/\partial T)_v ,\tag{2.8}$$

$$(\partial v/\partial T)_p = - (\partial S/\partial p)_T .\tag{2.9}$$

Chemical Potential

It is seen from equation (2.6) that the free energy of a system does not change if the temperature and pressure remain constant. Gibbs considered the conditions in a system of more than one component, such as in a solution, where there can be a change of free energy with a change in the number of moles of a component even though the pressure and temperature are constant. He defined a chemical potential of a component i as

$$\mu_i = (\partial G/\partial n_i)_{T, p, n_i} \tag{2.10}$$

or the change in free energy per unit change of the number of moles of the component while the temperature, pressure, and number of moles of the other components remain constant. The free-energy differential then has an added summation of the chemical potentials of all components, so that it is written in the form

$$dG = v\,dp - S\,dT + \Sigma(\partial G/\partial n_i)_{T, p, n}\,dn_i . \tag{2.11}$$

At constant pressure and temperature there can be a change in the free energy due to the chemical potential, because then

$$dG = \Sigma\mu_i\,dn_i . \tag{2.12}$$

In equilibrium, with the number of moles of the components constant, this expression would be equal to zero.

Consider a single, pure substance of n_1 moles. Since it occurs alone, $n_2 = 0$, $n_3 = 0$, etc. Then its chemical potential is

$$\mu_1 = (\partial G/\partial n_1)_{T, p} = G_1 . \tag{2.13}$$

This expression gives the chemical potential as one mole of the pure substance is added to the number of moles of the same substance at constant temperature and pressure. The increase of G of the system is therefore the value of G for one mole, that is, G_1. The chemical potential of a single pure substance is the free energy per mole of the substance. Note that in the symbolism of (1.14) and of (2.1) through (2.9) we are referring to one mole.

For an ideal gas at constant temperature, (2.6) becomes

$$dG = v\,dp = RT\,dp/p = RT\,d(\ln p) . \tag{2.14}$$

From the preceding paragraph it is apparent also that, for an ideal gas, μ_i is the free energy per mole at pressure p_i. Therefore,

$$\mu_i - \mu_i^0 = RT \ln (p_i/p_i^0) , \tag{2.15}$$

where the superscript 0 refers to a standard state.

The escaping tendency of a substance from a solution, a solid, a mixture, or a container at constant temperature can be characterized by its *fugacity*, defined in such a way that

$$d\mu = v\,dp = RT\,d(\ln y) , \tag{2.16}$$

21

$$\mu_i - \mu_i^0 = RT \ln (y_i/y_i^0) , \tag{2.17}$$

where y_i is the fugacity of the substance and y_i^0 is the fugacity of the standard state. One may define the standard state as one of unit fugacity, $y_i^0 = 1$, so that

$$\mu_i - \mu_i^0 = RT \ln y_i . \tag{2.18}$$

It is seen that for a pure gas at constant temperature

$$dG = d\mu = v \, dp$$

and that if the gas is ideal, we can write the differential form of (2.15),

$$d\mu = RT \, d(\ln p) , \tag{2.19}$$

which, when compared with (2.16), shows that the fugacity of an ideal gas is equal to its pressure.[1]

When vapor is in equilibrium with its liquid phase at the same temperature, its pressure is, by definition, equal to the vapor tension of the liquid. Then, according to equation (2.15), the two phases would be at the same level of free energy, $\mu_1 = \mu_2 = 0$. Also, since both temperature and pressure are constant, $dG = 0$, that is, there would be no change in the free energy of the system, even though heat may be exchanged and work performed. Thus the equilibrium may be defined as a free-energy equilibrium.

Vapor Pressure over Solutions

The vapor pressure of a component over a solution is a measure of the escaping tendency—the fugacity—of the component. It is equal to the partial pressure when the vapor behaves as an ideal gas. The chemical potential of a component in the solution must equal its chemical potential in the vapor phase, and from (2.18) it is seen that this is related to the fugacity. The partial vapor pressures are an indication of the chemical potentials of the components and, in the ideal case, provide an exact measure.

In a solution, we are dealing with two components, but in the usual case we are concerned with the vapor pressure of the solvent, e.g., the water-vapor pressure over a salt solution. The vapor pressure is, with unimportant exceptions, lower over the solution than over the pure solvent.

The fact that the water-vapor tension diminishes with increasing concentration and varies among the different solutes may be accepted on empirical grounds. Or, one may seek an explanation on the basis of molecular kinetics,

---

[1] The expression may be modified to apply to a non-ideal gas by considering that it approaches ideality at $p = 0$ and performing an integration from 0 to p to obtain an expression

$$RT \ln y = RT \ln p - \int_0^p a \, dp ,$$

where $a = v_{\text{ideal}} - v_{\text{real}} = (RT/p) - v$ and $v = (RT/p) - a$.

perhaps correlated with some other known property of the solution. It is logical to think that water molecules entangled with molecules of another substance which has an extremely low vapor tension would not leave the surface as freely as they would in the absence of these other molecules. The exact arrangement of the molecules and the cohesive forces between them in various solutions are not well known and a complete molecular-kinetic computation is not available. In the case of aqueous solutions of electrolytes the highly polar water molecules are attracted by the ions, especially the cations, which have been dissociated from the solute, and thus the fugacity of the water is reduced below that found for a non-electrolyte solution of the same molar concentration.

Solutions of most non-electrolytes behave as "ideal" solutions, defined as those in which the fugacity of each component is proportional to the mole fraction of that component in the solution. François Raoult first demonstrated this in 1886 on the basis of measurements of vapor pressure, showing that

$$\frac{p_A}{p_A{}^0} = \frac{n_A}{n_B + n_A}. \tag{2.20}$$

In this expression p_A is the vapor pressure of A above a solution of n_A moles of A and n_B moles of B while $p_A{}^0$ is the vapor pressure of pure A at the same temperature. The ratio $p_A/p_A{}^0$ is the *saturation ratio* which, multiplied by 100 would give the relative humidity (per cent) in a confined space over the solution. If the addition of B to pure A lowers the vapor pressure, the expression is conveniently written as

$$\frac{p_A}{p_A{}^0} = 1 - \frac{n_B}{n_B + n_A}, \tag{2.21}$$

where the second term is the commonly used mole fraction of solute.

Raoult's law for ideal solutions has to be modified for solutions of electrolytes. It is a frequent practice in physical chemistry in writing an expression for the vapor tension of a component in a solution of an electrolyte to relate it to some other property of the solution. For aqueous solutions the best approach, however, may be to use directly the empirically determined "water activity" which, for an ideal vapor like water, is equal to the saturation ratio $p_w/p_w{}^0$. The choice of property depends on what is available in published tables or, if no usable property is tabulated concerning the solution in question, one uses the relevant quantity which is most easily determined. The direct measurement of the vapor pressure may in some cases be the easiest. A related property that is often used is the *osmotic pressure*, the application of which to the problem at hand will now be discussed.

If a solution of B in A is separated from pure A by a membrane permeable only to A, an osmotic pressure is built up in the solution until an equilibrium is reached. At a given temperature the equilibrium state is determined by the concentration and some chemical properties of the solution.

In the equilibrium the chemical potential μ_A of A is the same on both sides of the membrane. In the solution two opposing effects balance each other to make the μ_A the same as in the pure A. One is due to the lowering of the vapor pressure and the other to the development of the osmotic pressure.

The chemical potential is the partial molar free energy and, at constant T, $d\mu = \bar{v}\, dp$, where $\bar{v}$ is the partial molar volume $[\bar{v}_A = (\partial V / \partial n_A)_{T,\, p,\, n_B}$, where V is total volume and n_A and n_B are the number of moles of A and B, respectively]. In the lowering of the vapor pressure we are concerned with an ideal gas, so the decrease in chemical potential may be written as

$$\mu_A{}^0 - \mu_A = \int_{p_A}^{p_A{}^0} RT\, d(\ln p) = RT \ln \frac{p_A{}^0}{p_A}, \qquad (2.22)$$

where the superscript 0 refers to pure A. The increase in chemical potential with the buildup of osmotic pressure from zero in the pure state to Π in the solution is

$$\mu - \mu_A{}^0 = \int_0^{\Pi} \bar{v}_A\, dp. \qquad (2.23)$$

If the volume of the solution is assumed not to change with pressure, we may set the equilibrium in balance by equating (2.22) and (2.23) as

$$\Pi = \frac{RT}{\bar{v}_A} \ln \frac{p_A{}^0}{p_A}. \qquad (2.24)$$

The inference of this equation is that the osmotic pressure is equivalent to an external pressure that would have to be applied to raise the vapor pressure of solvent A to that of pure A.

In three papers in 1885, 1886, and 1887, van't Hoff gave his classical demonstration that osmotic pressure, volume, and temperature are related in a manner analogous to the p,v,T relations of an ideal gas. His expression was of the form

$$\Pi = i\, n_B \frac{RT}{V}, \qquad (2.25)$$

where i was introduced as a factor, related to the dissociation of the ions in an electrolyte solution, which was needed to make the equation balance. Note that van't Hoff used the total volume, V.

Modern treatments use a coefficient applicable to the partial molar volume and some quantity expressing the concentration, such as

$$\Pi = \nu\varphi \frac{RT}{\bar{v}_A} \frac{n_B}{n_A} = \nu\varphi \frac{RT}{\bar{v}_A} \frac{f m_A}{1000}, \qquad (2.26)$$

where f is the molality (moles of B per kg of A) and ν is the number of moles of ions formed by the ionization of one mole of solute. The coefficient φ is the *molal osmotic coefficient*. Substitution of Π into (2.24) results in

$$\ln \frac{p_A}{p_A{}^0} = -\nu\varphi \frac{n_B}{n_A} = -\nu\varphi f m_A 10^{-3}. \qquad (2.27)$$

When the vapor pressures are replaced by the fugacities, this becomes the defining equation for φ, but once φ is known, (2.27) gives the relation of vapor pressure over the solution to that over pure A.

The osmotic coefficient is usually computed from the *activity coefficient* of the solute by means of relationships which will not be repeated here. The ν factor is based on the ions available for complete ionization. For a 1:1 electrolyte such as NaCl, $\nu \equiv 2$, for MgCl$_2$, $\nu \equiv 3$, for Al$_2$(SO$_4$)$_3 \equiv 5$, etc. Values of the osmotic coefficients for a few electrolytes are given in Table 2.1. Activity and

TABLE 2.1

OSMOTIC COEFFICIENTS OF SOME ELECTROLYTES AT 25° C

(Abridged from Robinson and Stokes, 1959)

MOLALITY f	NaCl $\nu \equiv 2$	MgCl$_2$ $\nu \equiv 3$	(NH$_4$)$_2$SO$_4$ $\nu \equiv 3$	Ca(NO$_3$)$_2$ $\nu \equiv 3$	Al$_2$(SO$_4$)$_3$ $\nu \equiv 5$
0.1........	0.932	0.861	0.767	0.827	0.420
0.2........	0.925	0.877	.731	0.819	.390
0.4........	0.920	0.919	.690	0.821	.421
0.6........	0.923	0.976	.667	0.831	.545
0.8........	0.929	1.036	.652	0.843	.718
1.0........	0.936	1.108	.640	0.859	0.922
1.2........	0.943	1.184	.632	0.879	....
1.6........	0.962	1.347	.624	0.917	....
2.0........	0.983	1.523	.623	0.953	....
2.5........	1.013	1.762	.626	1.001	....
3.0........	1.045	2.010	.635	1.051	....
3.5........	1.080	2.264	.647	1.103	....
4.0........	1.116	2.521	.660	1.157	....
4.5........	1.153	2.783	.673	1.210	....
5.0........	1.192	3.048	.686	1.263	....
5.5........	1.231	....	0.699	1.313	....
6.0........	1.272	....	....	1.361	....

osmotic coefficients have been tabulated by Robinson and Stokes (1959), Harned and Owen (1950), and by Hamer (1959).

Our interest in the solution effect in cloud physics relates to the fact that many cloud droplets start by condensation on soluble nuclei to form solution droplets. A curvature effect on the vapor tension of small droplets will be taken up in a later section. The equations in this section apply to solutions in bulk.

Latent Heat and Phase Change

A change of phase is accomplished at constant temperature and pressure while heat is exchanged between the system and its environment. The heat thus exchanged is referred to as the latent heat because it is revealed as an inherent characteristic of the substance only during phase change, in contrast to the sensible heat, the changes of which are continuously measured through the change in temperature.

Consider a p,v diagram as in Figure 2.1. For an ideal gas where no phase change is involved, the isotherms on the diagram would be in the form of perfect rectangular hyperbolas. In the figure, however, a region of the diagram in which a change of phase occurs is shown under the dashed curve. Let us suppose we are compressing the gas isothermally along T_3, removing heat at a rate just fast enough to keep the temperature constant. As the compression goes on, the pressure must be continually increased, as shown by the upward trend of the isotherm. The heat that is exchanged is given as

$$\delta'Q = T \, dS = p \, dv \tag{2.28}$$

for an isothermal process. But for an ideal gas at constant temperature,

$$p \, dv = -v \, dp = -RT \, d(\ln p) . \tag{2.29}$$

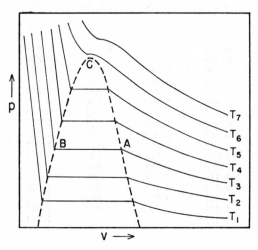

Fig. 2.1.—Diagram in the p,v plane showing various isotherms and the transition zone for phase change under the dashed curve.

This quantity is the same as the decrease in free energy per mole at the pressure p. For a 1/1000th increase in pressure at 290° K a computation shows that in order to preserve the temperature, 2.4×10^7 ergs or 0.576 calories per mole must be removed. Then suddenly at the point A, without any increase in pressure at all, the volume decreases to a small fraction of its value, and a very large amount of heat is released. For water vapor at 20° C (293° K) this heat amounts to 586 calories per gram or 10,584 calories per mole (18 grams). The heat released is the latent heat of condensation, L. From (2.31) it is seen that it is proportional to the difference in volume and to the difference in entropy between the points A and B. To the right of point A there is only vapor and to the left of B only liquid. The change in volume is the difference between the

26

64238

volume occupied by a mole of vapor at *A* and by a mole of liquid (at *B*). The volume of water vapor is about 25,000 times that of liquid water.

During the phase change, since the temperature and pressure are constant, the free energy is constant, $dG = 0$. The entropy and the enthalpy decrease within the system as heat is taken out, while the work content or Helmholtz free energy increases. Since the transition is accomplished at constant temperature and pressure, it is seen to be a spontaneous action of the gas molecules occurring at the condensation point.

The process is completely reversible, that is, it takes up the same amount of heat in changing from liquid to vapor as it released in condensation at the given temperature. Thus the heat of vaporization is equal to the heat of condensation but with the exchange in the opposite direction.

Other volume changes at constant pressure and temperature occur with melting, which exchanges the latent heat of fusion, and with sublimation, which is the passage from the solid directly to the vapor phase.[2] Within the range of measurement the heat of sublimation is numerically but not thermodynamically equal to the sum of the heats of fusion and of vaporization.

Values of the latent heats are found in physical, chemical, and meteorological tables. For water, the variations with temperature are small but nevertheless significant.

In considering the dashed curve of Figure 2.1, one notes that it has a maximum at *C*, called the critical point. The isotherm passing through this point corresponds to the critical temperature, defined as the temperature above which the gas cannot be liquefied. Gases below their critical temperature are often called vapor, especially, of course, water vapor which has its critical temperature at 647° K and its critical point at a pressure of 217.7 atmospheres and volume of 45 cm³ per mole. In this region we no longer have ideal-gas behavior, as shown by the non-hyperbolic character of the isotherms.

Since a gas or vapor has three variables of state (p, v, and T), a phase diagram such as in Figure 2.1 can be extended in a third dimension involving T as a coordinate. In the p,T plane the diagram for water in the atmospheric range is represented in Figure 2.2. The projection on this plane of the various surfaces of constant volume would produce lines (not shown) also running from lower left to upper right. Of greater interest is the transition zone included within the dashed curve of the previous figure (2.1). With the pressure and temperature constant in the transition, this zone forms a surface perpendicular to a p,T plane, so it intersects all such planes along the same curved line, shown in Figure 2.2. The transitions between liquid and ice and between vapor and ice are also shown in the figure. What the curve shows is the equilibrium pressure between pure water and its vapor at various temperatures, that is, the vapor

[2] The term sublimation is widely used to describe the reverse process, that is, the change from vapor to solid. McDonald (1958) gives logical reasons for using the word deposition for this reverse process.

27

MERNER - PFEIFFER LIBRARY
TENNESSEE WESLEYAN COLLEGE
ATHENS, TN. 37303

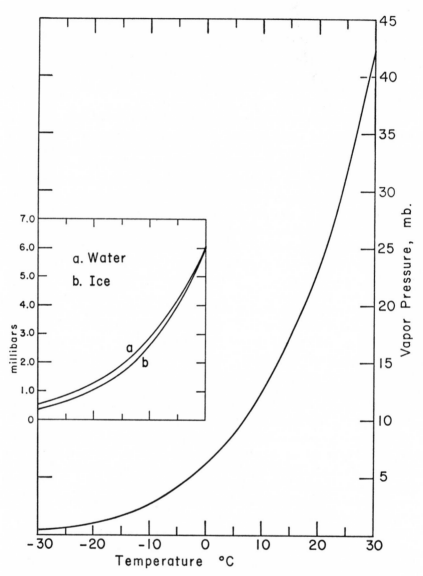

FIG. 2.2.—Phase diagram for water in the p,T plane *Inset:* vapor-ice equilibrium (lower curve) and vapor-liquid equilibrium (upper curve) at subfreezing temperatures.

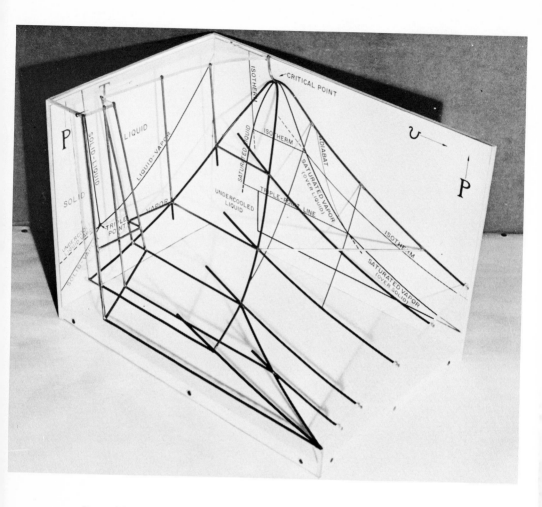

PLATE 2.1.—Three-dimensional phase model for water. (After McDonald, 1962.)

tension or saturation vapor pressure which, as we have seen, depends only on the temperature.

Below 0° C the curve has two branches, one for the vapor tension of ice and the other for the vapor tension of undercooled liquid. It is to be noted that the vapor pressure over undercooled water is greater than that over ice at the same temperature.

A three-dimensional model constructed of wire rods from a design by Mc-Donald (1962) is pictured in Plate 1. The isotherms form the skeleton of a surface representing the equilibrium state of each phase, which may be defined by the appropriate equation of state. The transition zone is in a surface that slopes upward and back toward increasing pressures and temperatures.

The ice $\rightleftharpoons$ liquid transition is represented by a slightly sloping wall that juts in from the p,T plane at the melting temperature. The direction of projection of this surface is peculiar to water, signifying the decrease in volume as ice melts.

Except for the essentially imperceptible 0.0075° displacement of the melting point, the equilibrium values are the same as they would be if the atmosphere were not present. This fact is in agreement with Dalton's law.

For pure water vapor without other atmospheric gases present, the triple point with the three phases in equilibrium at the intersection of the wall and the transition surface occurs at a temperature of 0.0075° C. In the presence of the total atmosphere 0° C is, by definition, the melting point. At the boiling point the vapor tension is the same as the atmospheric pressure and, again by definition, 100° C is the boiling point of pure water at the standard atmospheric pressure of 1.01325 bars.

The passage of vapor to ice is represented by two parts to the transition surface at temperatures below the melting point: a lower "lip" and an upper one. The lower one represents the transition at equilibrium between ice and the vapor while the upper one represents the transition from vapor to undercooled water. The latter extends into the ice wall. It should be remembered, however, that this transition is not reversible; water has no fixed freezing point, but 0° C is the melting point. When undercooled water freezes, its temperature goes up to 0° C and remains there until all the water is frozen—assuming that we are here talking about small amounts as in drops—then if it cools and makes the transition to vapor it does so directly on the lower "lip" of the transition surface.

If for some reason the equilibrium between liquid and vapor occurs at some degree of *supersaturation* with respect to a flat surface of pure water, the transition would be shifted upward and to the left in the model. This state is suggested in the model by indefinite small extensions of the isotherms above the transition surface. Equilibrium at subsaturation would shift the transition to the right and downward. While in these two cases the free energy would be constant, as it is on the reference transition-zone isotherms, the free-energy levels would be different. The supersaturation represents a metastable state which, as shown in a later section, can be a state of stable equilibrium for small droplets.[3]

The heat of fusion that must be extracted for freezing of undercooled water is less than that required for water at 0° C. The difference is about equal to the amount of sensible heat that would have to be removed to cool the same mass of water in the solid form from 0° C to the original undercooled temperature, or that would have to be added in the reverse process. In other words, this much heat having been removed already in the undercooling, it does not need to be removed in order to freeze the water. But in the freezing process the water-ice mixture goes to 0° C, so in order to cool it back down to the original temperature the heat amounting to the above-mentioned difference would have to be extracted. Thus to freeze undercooled water with no final change in the temperature would require a heat removal about equivalent to the latent heat of fusion at 0° C.

The Clapeyron-Clausius Equation

The slopes of the curves of Figure 2.2 can be determined by a relation originally put forward by Clapeyron in 1834 and placed on a more sound thermodynamic basis some 30 years later by Clausius. The equation involves the equilibrium between the two phases α and β of a one-component system. In equilibrium the free energies per mole in the two phases are equal.

To obtain the slope we consider two points very close together on the curve, namely, p,T with free energy $G_\alpha = G_\beta$, and $p + dp$, $T + dT$ with free energy $G_\alpha + dG_\alpha = G_\beta + dG_\beta$. It follows that $dG_\alpha = dG_\beta$. Substituting from (2.6), we have

$$v_\alpha \, dp - S_\alpha \, dT = v_\beta \, dp - S_\beta \, dT \, . \qquad (2.30)$$

If $v_\beta > v_\alpha$,

$$\frac{dp}{dT} = \frac{S_\beta - S_\alpha}{v_\beta - v_\alpha} = \frac{\Delta S}{\Delta v} \, . \qquad (2.31)$$

But $\Delta S = \Delta Q/T = L/T$, where ΔQ is the heat exchanged in the phase transformation, or the latent heat per mole, L. We are then led to the Clapeyron-Clausius equation in the form

$$\frac{dp}{dT} = \frac{L}{T\Delta v} \, . \qquad (2.32)$$

By applying appropriate values of L, one can use the equation for melting-fusion, vaporization-condensation or sublimation-deposition. The sign of Δv must also be known. For example, for the melting of water the decrease in volume $(-\Delta v)$ would correspond to a decreasing pressure with increasing temperature. For vaporization and sublimation all terms would be positive.

[3] Certain interesting properties of water at very high pressures and some other unstable and metastable states are not of interest in the atmosphere. The reader is referred to physical chemistry textbooks or encyclopedia articles.

In the transitions vapor $\rightleftharpoons$ liquid or vapor $\rightleftharpoons$ solid some simplifications can be made in the equation, especially if the vapor behaves as an ideal gas. In the case of water, v_β is 25,000 v_α, so $v_\beta - v_\alpha \sim v_\beta = RT/p$, and we may write

$$\frac{dp}{dT} = \frac{L}{Tv_\beta} = \frac{Lp}{RT^2} \qquad (2.33)$$

or

$$\frac{d(\ln p)}{dT} = \frac{L}{RT^2}, \qquad (2.34)$$

which is the Clapeyron-Clausius equation for water and other substances that vaporize to essentially ideal gases at moderate temperatures, governing the vaporization-condensation and sublimation-deposition processes.

The equation was derived at a given temperature, that is, at a given point on the p,T curve, but L varies as $T\Delta S$. The variation with temperature has been determined experimentally and is represented by empirical formulas. One normally obtains the values from tables.

Latent heats are given in most tables in calories per gram. Of course the calories have to be multiplied by the mechanical equivalent of heat to obtain the work units used here, and the conversion to unit mass is accomplished by using the molecular weight where appropriate in the preceding equations. The v's become specific volume, a, and R becomes R/m.

Effects of Surface Free Energy

In considering phase equilibria it is necessary to realize that interfacial surfaces separate the phases. The equilibrium is between the molecules of the two phases at such a surface, but other surface energies or forces may affect the equilibrium. Prominent among these is the surface free energy.

When the area of an interface such as the surface of a liquid with its vapor is increased, molecules from the interior of the liquid have to be brought to the surface. In this process work must be done against the cohesive forces in the liquid. The molecules in the interior are surrounded on all sides by the field of force of neighboring molecules while the molecules at the surfaces have neighbors of the same phase on one side and only the widely scattered molecules of the vapor on the other. The surface layer has a higher free energy than the main body of the liquid.

The *surface tension* is defined as the work per unit area done in extending the surface of the liquid. The surface free energy is increased by an amount equivalent to the amount of work done,[4] so the surface tension may be expressed as

$$\sigma = dG/dA . \qquad (2.35)$$

---

[4] Thermodynamically, since this is a change in work potential, it is the Helmholtz free energy that is involved. Although the equivalence is accepted as fact, it cannot be thermodynamically demonstrated in a rigid manner.

Dimensionally, work or energy per unit area is the same as force per unit length. Physically, in a soap bubble this is seen as the force pulling across a circumference line; $\sigma = \text{force}/2\pi r$ on a sphere, and, since pressure is force per unit area, $\sigma = PA'/2\pi r$, where A' is the area of a cross-section of the sphere through a circumference and P is the pressure exerted in performing the work of extending the surface. This simplifies to $\sigma = rP/2$.

The concept of surface tension as force per unit length or as a pressure times a length has limited use. It is best to think of surface tension as the free energy per unit area of surface or the *specific* surface free energy.

Vapor Pressure over Pure Droplets

It is found from measurement that the vapor tension of a concave surface of a liquid is lower than that of a flat surface. Convex surfaces show higher vapor tensions than flat surfaces. Droplets therefore require a certain degree of supersaturation in the ambient space in order to remain in equilibrium. Physically, one might indeed expect the escaping tendency of molecules from a spherical droplet to be greater the smaller the droplet because the component of the binding force in the tangential direction is diminished. Lord Kelvin (William Thomson) first deduced the relation which today can be presented in more modern thermodynamic form from the point of view of the surface free energy.

In the supersaturated state the transition surface lies above that depicted in Plate 2.1, each isotherm being at a higher pressure and therefore at a higher free energy. The elevation of the free energy above that for a plane surface of pure water depends on the saturation ratio and on the surface free energy.

The problem is approached by considering a system initially consisting of vapor at supersaturation, with a saturation ratio S which changes to a final state, and forming from the vapor a single embryonic droplet of radius r containing g molecules. Note that

$$g = \tfrac{4}{3}\pi r^3 Z_L ,$$

where Z_L is the number of molecules in a cubic centimeter of the liquid, and that $Z_L = \mathfrak{N}\rho_L/m_w$, where $\mathfrak{N}$ is Avogadro's number (6.022×10^{23}). If molecules are considered in place of moles, $\mathfrak{N}k$ is substituted for R, where k is Boltzmann's constant (1.381×10^{16} erg °K^{-1} molecule^{-1}, or effectively the gas constant per molecule). The free-energy differential dG of equation (2.14) becomes

$$dG = g\,kT\,d(\ln p) = \tfrac{4}{3}\pi r^3 Z_L kT\,d(\ln p) . \tag{2.36}$$

As g molecules go from the vapor at pressure p to the liquid at tension p_0 the change in chemical potential or free energy of the system is given, after the manner of equation (2.15), as

$$\mu_{p_0} - \mu_p = \int_p^{p_0} g\,kT\,d(\ln p) = g\,kT \ln \frac{p_0}{p} \tag{2.37}$$

$$\mu_{p_0} - \mu_p = -\tfrac{4}{3}\pi r^3 Z_L kT \ln \frac{p}{p_0} = -\tfrac{4}{3}\pi r^3 Z_L kT \ln S, \qquad (2.38)$$

where S is now the saturation ratio. As the transition is made from vapor only to an embryo droplet of radius r, the surface free energy goes from zero to $4\pi r^2 \sigma$. The elevation of the total free energy of the system of g molecules in the presence of these two effects is therefore their sum, or

$$\Delta G = 4\pi r^2 \sigma - \tfrac{4}{3}\pi r^3 Z_L kT \ln S . \qquad (2.39)$$

The significance of this expression may best be appreciated by plotting ΔG against r for several values of S at constant temperature, and therefore also constant σ, as in Figure 2.3. It is seen that each curve has a maximum. These curves represent a free-energy barrier to the growth of embryos at the given supersaturations. The free-energy level must be at least as high as the peak of the curve in order for the embryo to grow. The radius at which the maximum occurs is known as the *critical* radius, r^*. If the radius becomes larger with the same saturation ratio, growth is assured, for then the vapor would be at a higher free energy than the liquid. The less the supersaturation, the greater the barrier and the larger r^* must be.

If one accepts the rule that a process is spontaneous if the free energy decreases, then one sees that if the system can reach the peak of its ΔG curve, a slight addition of molecules to the droplet will allow it to grow irreversibly, sliding down to the right of its peak as plotted in Figure 2.3. From the curves it is apparent that the higher the supersaturation the lower the free-energy peak or barrier.

Since the critical radius is at the maximum of the curve for each S, it is obtained by differentiating (2.39) with respect to r and setting the derivative equal to zero.

$$\left[\frac{\partial (\Delta G)}{\partial r} \right]_{T, S} = 8\pi r \sigma - 4\pi r^2 Z_L kT \ln S, \qquad (2.40)$$

which is zero when the two terms on the right are equal or when

$$\ln S = \frac{2\sigma}{Z_L kT r}; \qquad (2.41)$$

and, since at this point r is critical,

$$r^* = \frac{2\sigma}{Z_L kT \ln S} = \frac{2 m_w \sigma}{\rho_L RT \ln S}, \qquad (2.42)$$

which is the Kelvin equation. The second derivative is negative at r^*, indicating a maximum.

A plot of S against r^* is given in Figure 2.4 (upper curve marked "Pure").

33

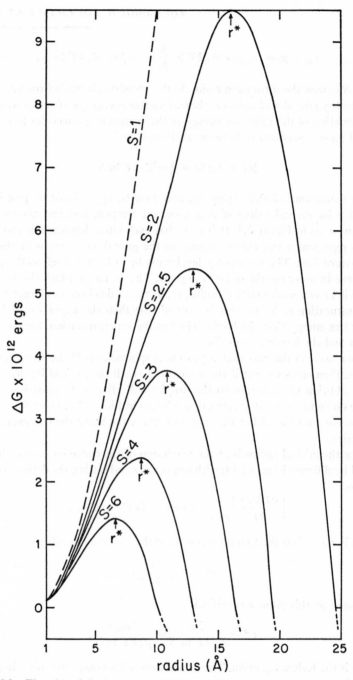

Fig. 2.3.—Elevation of the free energy ΔG above equilibrium for a plane surface of pure water plotted against radius for various values of the saturation ratio, S.

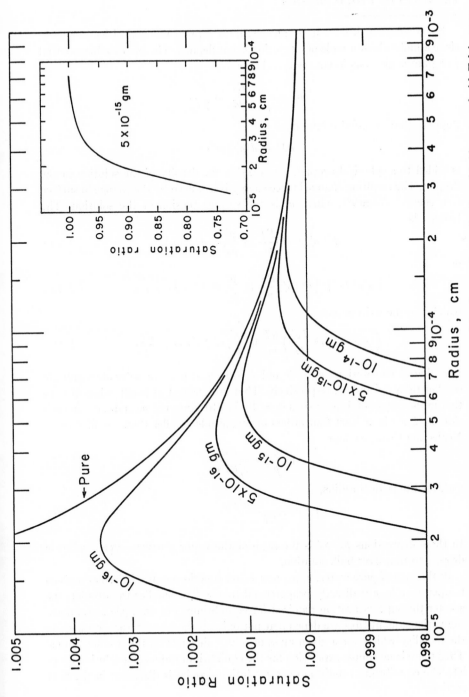

Fig. 2.4.—Curves of equilibrium saturation ratio of water droplets containing the stated mass of sodium chloride compared with Kelvin curve for pure water droplets. *Inset:* curve for 5×10^{-15} g NaCl on a compressed scale extended to the droplet size at which the given amount of NaCl would form a saturated salt solution in the droplet. All computations are made for a temperature of 25° C, but the values are very nearly the same at other atmospheric temperatures.

For a solution droplet it is convenient to express the quantities in grams, since a molecule or a mole of a solution is not defined. The increase in potential per gram is given as in equation (2.15) by

$$\mu' - \mu_0' = \frac{RT}{m} \ln \frac{p'}{p_0'}.$$ (2.43)

Suppose that an infinitesimal mass

$$dM = d(\rho'_{L} \tfrac{4}{3}\pi r^3)$$

is added to a spherical droplet, where ρ_L' is the density of the solution in the droplet. The resulting change in free energy is balanced by the change in surface free energy, $d(4\pi\sigma'r^2)$, where σ' is the surface tension of the solution. The balance is

$$d \left(\rho'_{L} \tfrac{4}{3}\pi r^3 \right) \frac{RT}{m_w} \ln \frac{p'}{p_0'} = d(4\pi\sigma'r^2)$$ (2.44)

or

$$\left(\rho'_{L} r^2 dr + \tfrac{1}{3} r^3 d\rho'_{L} \right) \frac{RT}{m_w} \ln \frac{p'}{p_0'} = 2\sigma'r \, dr + r^2 d\sigma',$$ (2.45)

which may be written as

$$\left(\rho'_{L} r^2 + \tfrac{1}{3} r^3 \frac{d\rho'_{L}}{dr} \right) \frac{RT}{m_w} \ln \frac{p'}{p_0'} = 2\sigma'r + r^2 \frac{d\sigma'}{dr}.$$ (2.46)

For all aqueous solutions, $d\rho_L'/dr$ and $d\sigma'/dr$ are at least an order of magnitude smaller than ρ_L' and σ', respectively. They are greatest at small values of r, so that in the ranges of droplet sizes ($r = 10^{-3}$ to 10^{-6} cm) the second term on each side of (2.46) is at least four orders of magnitude smaller than the first term. Neglecting them, we find

$$\ln \frac{p_r'}{p_0'} = \frac{2\sigma' m_w}{\rho'_L RT r}$$ (2.47)

and, for the critical radius,

$$r^* = \frac{2\sigma' m_w}{\rho'_L RT \ln(p_r'/p_0')}.$$ (2.48)

In these expressions p_r'/p_0' is the ratio of the vapor pressure over a solution droplet to that over bulk solution.

In the case of pure water, σ, ρ_L, and p_0 are functions only of T, so at a given temperature $\ln p$ is directly proportional to $1/r$. The smaller the droplet, the greater the supersaturation required for equilibrium. For an aqueous solution, the various quantitites are dependent not only on the temperature but also on the molality which, for a given mass of solute, depends on the droplet radius. Thus there is no simple expression for the equilibrium radius of a solution droplet. The equilibrium condition over a solution droplet is discussed in the next section.

36

Equilibrium over Solution Droplets

The two expressions (2.27) and (2.42) can be combined into a single expression for the equilibrium over a solution droplet. The vapor tensions involved are those of four different kinds of liquid surface in contact with the vapor: p_0, plane surface of pure water; p_0', plane surface of aqueous solution; p_r, pure water droplet of radius r; p_r', solution droplet of radius r. What is wanted in the combined equation is p_r'/p_0 the vapor saturation ratio for a solution droplet with reference to a plane, pure water surface. We simply write

$$\frac{p_r'}{p_0} = \frac{p_0'}{p_0} \cdot \frac{p_r'}{p_0'} \tag{2.49}$$

and substitute for the two ratios on the right from equations (2.27) and (2.47) to obtain

$$\frac{p_r'}{p_0} = \exp\left(\frac{2m_w\sigma'}{\rho_L'RTr} - \frac{i\,f\,m_w}{1000}\right) \tag{2.50}$$

or

$$\ln\frac{p_r'}{p_0} = \frac{2m_w\sigma'}{\rho_L'RTr} - \frac{i\,f\,m_w}{1000}, \tag{2.51}$$

where i designates $\nu\varphi$ (not the same i that van't Hoff used in equation [2.25]).

Equation (2.50) will be greater or less than 1 depending on whether the solution effect or the curvature term predominates. In other words, solution droplets of high concentration of solute can be in equilibrium at subsaturation. For a given drop the molality varies as r^{-3} and the Kelvin pressure varies as r^{-1}. Very small droplets of solution counteract the Kelvin effect and are at equilibrium at low relative humidities.

Equilibrium curves as a function of radius for droplets containing a given nucleus at a given temperature are plotted in Figure 2.4. The positions of the curves vary only very slightly through a wide range of atmospheric temperatures. For comparison, a portion of the curve for a pure water droplet obeying the Kelvin equation is given. It is obvious that with an adequate supply of suitable nuclei, droplets would not be expected to form on embryos consisting of molecules of pure water.

Nucleation of cloud droplets is discussed in detail in chapter 3, and the full meaning of the curves of Figure 2.4 is emphasized. The curves are sometimes referred to as Köhler curves after the Swedish atmospheric chemist who first developed them for atmospheric conditions in 1926.

GENERAL WORKS

DUFOUR, L., and DEFAY, R. (1963), *Thermodynamics of Clouds* (New York: Academic Press).

FLETCHER, N. H. (1962), *The Physics of Rainclouds* (Cambridge: Cambridge University Press).

MOORE, W. J. (1962), *Physical Chemistry* (3d ed.; New York: Prentice-Hall).

REFERENCES

GIBBS, J. W. (1875), *Trans. Conn. Acad.*, **3**, 108, or *Collected Works* (New York: Longmans, Green and Co. 1928).

HAMER, W. J., ed. (1959), *The Structure of Electrolytic Solutions* (New York: John Wiley and Sons).

HARNED, H. L., and OWEN, B. B. (1950), *Physical Chemistry of Electrolytic Solutions*, Amer. Chem. Soc. Monogr. No. 95.

KÖHLER, H. (1926), *Med. Statens Met.-Hydrogr. Anstalt* (Stockholm), **3**, No. 8.

MCDONALD, J. E. (1958), *J. Met.*, **15**, 245.

———. (1962), *Bull. Amer. Met. Soc.*, **43**, 292.

ROBINSON, R. A., and STOKES, R. H. (1959), *Electrolyte Solutions* (London: Butterworth's).

VAN'T HOFF, J. H. (1885), *Arch. Néerlandaises*, Année 1885.

———. (1886), *Sveriges Akad. Handlingar*, **21**, No. 12, 43.

———. (1887), *Zeitschr. Phys. Chem.*, **1**, 481.

3

NUCLEATION PROCESSES

In the previous chapter equilibrium states between phases were discussed with only general implications concerning the ease of passage from one phase to the other. It was implied that a certain infinitesimal excess of the ambient vapor pressure over the equilibrium value would cause condensation to occur. The common observation of marked undercooling before freezing begins in clouds was mentioned without specifying determining factors.

Because of the solution effect, it was noted that droplets will grow easily without appreciable supersaturation if they are formed on soluble particles. Nothing was said, however, about the possible role of insoluble particles; and nothing was said about particles which might serve as nuclei for freezing. Since electrolyte solutions generally have lower freezing temperatures than pure water, one would expect that the nuclei which are most favorable for condensation would not be effective for freezing. In fact, the formation of ice crystals and the freezing of droplets are in a much lower realm of probability than in the case of condensation, and these processes require a different set of nucleating particles which are less abundant and less active than condensation nuclei.

At high supersaturations the number of molecules contained in an embryo droplet of pure water in equilibrium with the vapor can be quite small. Some values are given in Table 3.1.

The Two Kinds of Nucleation

When a suitable surface is present, nucleation (condensation, ice deposition, freezing) will occur on it. Such surfaces may be foreign particles (condensation nuclei, freezing nuclei) or larger surfaces of foreign substances or of the same substance. Such nucleation is called *heterogeneous*. In the absence of suitable surfaces the nucleation would have to take place on embryos made up of relatively compact collections of a few molecules of the substance. Such nucleation is called *homogeneous*. Figure 2.4 shows that homogeneous nucleation of pure water vapor to droplets requires large supersaturations until the droplets have grown to sizes where the curvature effect is negligible.

The only way embryos for homogeneous nucleation can be formed is by a chance succession of collisions which results in the sticking together of numbers of molecules. Such an event is improbable because the total binding force exerted by small aggregates of molecules upon their surface members is too small to counteract the thermal motions tending to carry them away. There may not be enough inside molecules to attract and hold an outer molecule. If the supersaturation is very great, molecules from the vapor phase are driven *to* the embryos faster than molecules are lost *from* them. But first there must be embryos.

TABLE 3.1

RADII AND NUMBER OF MOLECULES IN
DROPLETS OF PURE WATER IN EQUI-
LIBRIUM WITH THE VAPOR AT 0° C*

Saturation Ratio S	Equilibrium Radius $r*$ (angstroms)	Number of Molecules $g*$
1.............	∞	....
2.............	17.3	714
2.5..........	13.1	309
3.............	10.9	164
4.............	8.7	89
5.............	6.7	41

* The $r*$ is from Kelvin equation; $g* = 4\pi r*^3 \Re \rho / 3m_w$, where $\Re$ is Avogadro's number.

Homogeneous Nucleation, Vapor to Liquid

The problem of homogeneous nucleation of liquid droplets can be treated analytically from the point of view of the probability of occurrence of embryos under different temperatures and supersaturations. The question concerns the surmounting of the free-energy barrier represented by the maximum of the curve for the elevation of the free energy above that for a plane surface of pure water, as expressed by equation (2.39), repeated here:

$$\Delta G = 4\pi r^2 \sigma - \tfrac{4}{3}\pi r^3 Z_L kT \ln S . \qquad (3.1)$$

The peak of the barrier is at the equilibrium or critical radius $r*$ as given by the Kelvin equation (2.42). The substitution of the Kelvin equation for r in (3.1) gives the free-energy peak as

$$\Delta G* = \frac{16\pi\sigma^3}{3(Z_L kT \ln S)^2} . \qquad (3.2)$$

The expression was used in a slightly different form by F. J. M. Farley in a notable paper in 1952 (see eq. [3.3]). From it Farley derived the equilibrium condition as represented by the Kelvin equation, then developed a theory of condensation on embryos of the pure liquid—homogeneous nucleation.

Much of the basic theory in Farley's paper may be found in J. Frenkel's book, *Kinetic Theory of Liquids* (1946). The theory of condensation in terms of kinetic arguments was earlier considered by Becker and Döring (1935) as an improvement on a thermodynamic approach by Volmer and Weber (1926). Recently McDonald (1963) gave a clarification of this complicated problem in terms of physical arguments.

Theory of Embryo Growth

The physics of the growth and evaporation of embryos is on a molecular scale, so it is convenient to denote embryos in terms of the number of molecules they contain. An embryo is considered as being made up of g molecules. Thus, $g \sim r^3$ and $r^2 \sim g^{2/3}$, and we may write for the surface free energy as used in (3.1), $4\pi r^2 \sigma = \beta g^{2/3}$, where β is a constant for a given value of σ (any given temperature). Equation (3.1) may then be written as

$$\Delta G = \beta g^{2/3} - g \, kT \ln S . \qquad (3.3)$$

In Farley's treatment referred to in the previous section, a system is considered consisting of N_1 individual molecules some of which are in N_g embryos containing g molecules each. Farley introduced a distribution of embryos, according to the Boltzmann law, related to states of free energy in the form

$$N_g = N_1 \exp(- \Delta G / kT) = N_1 \exp \left(g \, \ln S - \frac{\beta \, g^{2/3}}{kT} \right). \qquad (3.4)$$

For small g and very large g this distribution departs from reality, but it may be considered appropriate in the ranges not too far from r^*, corresponding to g^*. For example, when $g = 1$, N_g should be identical with N_1, but the equation does not produce this identity. By definition, N_g cannot exceed N_1, but the plots in Figure 3.1 show that at large values of g the equation allows N_g to be greater than N_1. At the lowest end of the embryo spectrum the distribution would approach that defined by molecular-energy states; furthermore the specific surface free energy contained in the β term could no longer be defined.[1] At the upper end of the spectrum the distribution must cut off where the accumulated value of the product gN_g approaches N_1.

It is seen by differentiating (3.4) at constant T, N_1, and S ($S > 1$), that N_g has a minimum at

$$\tfrac{2}{3}\beta g^{-1/3} = kT \ln S = 8\pi \sigma r^2/3g . \qquad (3.5)$$

---

[1] A correction factor for the surface tension of small droplets as suggested by Tolman (1949), and its application to the problem at hand, will be considered subsequently.

41

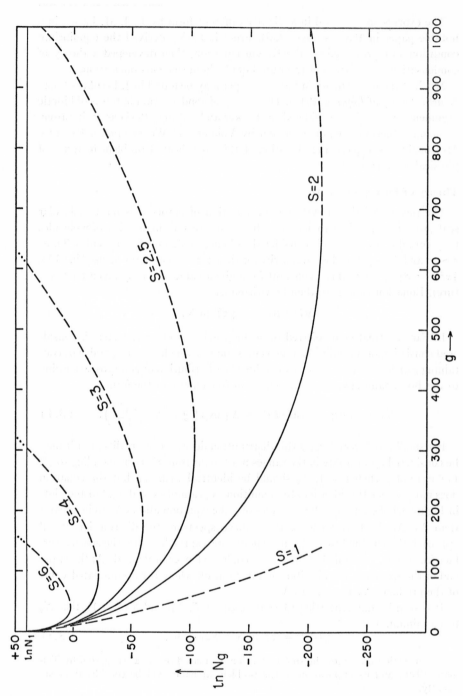

Fig. 3.1.—Boltzmann-type distribution of embryos according to the number of molecules, g, which they contain plotted for various saturation ratios, S.

But

$$g = \tfrac{4}{3}\pi r^3 (\mathfrak{N}\rho_L/m_w) \,,$$

and, since $\mathfrak{N}k = R$, we find the minimum is at

$$\ln S = 2m_w\sigma/(\rho_L RT\, r^*) \,, \tag{3.6}$$

which we recognize from the Kelvin equation as showing the minimum to be at the equilibrium radius or at g^*. It will be noted in Figure 3.1 that the greater the supersaturation, the farther upward and to the left would g^* be found.

Since g^* is at the free-energy barrier or maximum of Figure 2.3, it is clear that a distribution as in equation (3.4) which depends on a negative exponent of the free energy should exhibit a minimum number of embryos at the free-energy maximum. But physical reasoning does not suggest that with $g > g^*$ the embryos would be found in larger numbers the greater their size. As S increases into supersaturation the space contains many small embryos (as represented first by the curve for $S = 1$ in Fig. 3.1) and no large ones, and the equilibrium of (3.4) becomes increasingly difficult to achieve.

A way out of this dilemma can be found by treating the kinetics of the movement of molecules between each single g-step of the distribution. Going upward in the steps of g, one finds an increasing departure from the equilibrium distribution in the sense that there will always be more embryos of $g - 1$ molecules than of g molecules.

Farley examined the microscopic balancing conditions for *dynamic* equilibrium between embryos of size g, embryos of size $g - 1$, and the vapor, once the equilibrium distribution of (3.4) has been established.[2] If ξ_{g-1} represents the number of molecules striking an embryo of size $g - 1$ per second and ζ_g is the number of molecules evaporating from an embryo of size g per second, equilibrium demands that

$$\xi_{g-1}N_{g-1} - \zeta_g N_g = 0 \,. \tag{3.7}$$

In other words, no embryos pass from size $g - 1$ to size g.

The only way for the larger embryos to form is for the quantity in (3.7) not to equal zero. The quantity is I_g, the net number of embryos passing per second up the step from size $g - 1$ to size g. By the definition of equilibrium represented by (3.7) this move takes us out of the equilibrium. In this non-equilibrium state we shall designate the number of embryos by N'_{g-1}, N'_g, etc. Then

$$I_g = \xi_{g-1}N'_{g-1} - \zeta_g N_g' \,. \tag{3.8}$$

We wish to relate this to the point on the equilibrium distribution from which it branches. If we multiply and divide the first term on the right by N_{g-1} and the second by N_g, we have

$$I_g = \xi_{g-1}N_{g-1}\frac{N'_{g-1}}{N_{g-1}} - \zeta_g N_g \frac{N'_g}{N_g}\,; \tag{3.9}$$

---

[2] Note that the liberty is taken of speaking of the number of molecules in the embryo as its size.

then from (3.7) it is seen that the two products outside the two ratios are equal and

$$I_g = \xi_{g-1} N_{g-1} \left(\frac{N'_{g-1}}{N_{g-1}} - \frac{N_g'}{N_g} \right) = \xi_{g-1} N_{g-1} \left(-\Delta \frac{N_g'}{N_g} \right)$$

$$= - \xi_g N_g \frac{d}{dg} \left(\frac{N_g'}{N_g} \right). \tag{3.10}$$

A steady state is set up such that $I_g = I$, a constant independent of g. In this distribution the population N_g' is steady, but there is a net passage of embryos from small to larger sizes. The quantity I is the number of embryos forming per second from the vapor, passing through the distribution, and finally appearing as visible droplets at the rate I per second. The quantity I is called the *current*; N_g' is in the distribution that corresponds to the steady current I. As g increases N_g' falls more rapidly than N_g, passing through g^* while decreasing more slowly than at first. Through a small interval, ξ_g is relatively unchanging; therefore, from (3.10), with a constant current,

$$-\frac{d}{dg} \left(\frac{N_g'}{N_g} \right) = \frac{I}{\xi_g} \frac{1}{N_g} = \frac{\text{const}}{N_g}, \tag{3.11}$$

so although the ratio of N_g'/N_g becomes continually smaller, it does so at a diminishing rate as N_g increases beyond N_{g*}.

Farley integrates (3.11) by assuming that I and ξ are constant, writing, with subscripts dropped for convenience,

$$\frac{N'}{N} = -\frac{I}{\xi} \int \frac{1}{N} \, dg + A, \tag{3.12}$$

where A is a constant. The demonstration that $A = 0$ for large values of g will not be repeated here.

As $g \to 0$, $(N'/N) \to 1$ and the steady current becomes

$$I = \frac{\xi}{\displaystyle\int_0^\infty \frac{1}{N} \, dg}. \tag{3.13}$$

In the integral the value for N is substituted from (3.4), but with $1/N$ replaced by a function which approximates it in the important region around g^*, expressed as

$$\frac{1}{N} = \frac{1}{N^*} \exp \left[-\chi (g - g^*)^2 \right],$$

where

$$\chi = -\tfrac{1}{2} \frac{d^2}{dg^2} \left(\frac{\beta g^{2/3}}{kT} - g \ln S \right).$$

Upon taking this second derivative at g^* and r^*, one finds

$$\chi = 4\pi\sigma r^{*2}/9kTg^{*2}. \tag{3.14}$$

The integral becomes

$$\int_0^\infty \frac{1}{N}\, dg = \frac{1}{N^*}\int_0^\infty \exp[-\chi(g-g^*)^2]\, dg = \frac{1}{N^*}\left(\frac{\pi}{\chi}\right)^{1/2}$$

and

$$I = \frac{\xi}{\displaystyle\int_0^\infty \frac{1}{N}\, dg} = \xi N^* \left(\frac{\chi}{\pi}\right)^{1/2}. \tag{3.15}$$

The quantity ξ, the number of molecules striking an embryo of given size per second, can be expressed in terms of conventional kinetic theory of gases. As shown in standard texts, the average velocity of molecules of a gas at temperature T is

$$\bar{c} = (8RT/\pi m)^{1/2}, \tag{3.16}$$

where m is the molecular weight. Since a molecule moving in any plane can go with a component in any of the four directions, the number striking any square-centimeter surface per second would be $q = Z\bar{c}/4$, where Z is the number of molecules in a cubic centimeter. Substitution for $\bar{c}$ from (3.16) makes

$$q = Z(RT/2\pi m)^{1/2}.$$

But $Z = \mathfrak{N}\rho/m$ where $\mathfrak{N}$ is Avogadro's number, and since $\rho = pm/RT$ we find

$$q = \mathfrak{N}p(2\pi mRT)^{-1/2},$$

and, for an embryo of radius r composed of pure water of molecular weight m_w,

$$\xi = 4\pi r^2 \mathfrak{N} p(2\pi m_w RT)^{-1/2}. \tag{3.17}$$

A coefficient representing the fraction of those striking which actually stick may be inserted. For the present purpose this coefficient will be assumed to be one.

Substituting ξ from (3.17) into (3.15) as well as the value of χ from (3.14) yields

$$I = 4\pi r^2 p \mathfrak{N}(2\pi m_w RT)^{-1/2}\, \frac{2}{3}\, \frac{r^*}{g^*}\left(\frac{\sigma}{kT}\right)^{1/2} N^*, \tag{3.18}$$

but

$$g^* = \frac{\rho_L}{m_w}\, \mathfrak{N}\tfrac{4}{3}\pi r^{*3},$$

so

$$I = 2p(2\pi RT m_w)^{-1/2}\, \frac{m_w}{\rho_L}\left(\frac{\sigma}{kT}\right)^{1/2} N^*. \tag{3.19}$$

Now N^* is at the maximum determined from (3.1) when

$$8\pi r\sigma = 4\pi r^2 Z_L kT \ln S,$$

and, since from (3.1)

$$\Delta G = 4\pi r^2 \sigma - \tfrac{4}{3}\pi r^3 Z_L kT \ln S,$$

we find

$$\Delta G^* = 4\pi r^{*2}\sigma - \frac{r^*}{3} 8\pi r^*\sigma = \tfrac{4}{3}\pi r^{*2}\sigma$$

and

$$N^* = N_1 \exp\left(\frac{-\Delta G^*}{kT}\right) = N_1 \exp\left(\frac{-4\pi r^{*2}\sigma}{3kT}\right). \qquad (3.20)$$

We want to express I in terms of the number of droplets formed per second in a cubic centimeter. Then N_1 should be the total number of molecules in a cubic centimeter of the vapor at temperature T, or[3]

$$N_1 = \frac{\mathfrak{R}\rho_v}{m_w} = \frac{\mathfrak{R}p}{RT}$$

and

$$N^* = \frac{\mathfrak{R}p}{RT} \exp\left(\frac{-4\pi r^{*2}\sigma}{3kT}\right). \qquad (3.21)$$

Finally, with N^* substituted in (3.19),

$$
\begin{aligned}
I = 2p(2\pi RT m_w)^{-1/2} \frac{m_w}{\rho_L}\left(\frac{\sigma}{kT}\right)^{1/2} \frac{\mathfrak{R}p}{RT} \exp \frac{-4\pi r^{*2}\sigma}{3kT} \\
= f(p,T)N_1 \exp\frac{-\Delta G^*}{kT}.
\end{aligned} \qquad (3.22)
$$

Then one may introduce molecular constants with the mass of a molecule $\mathfrak{m} = m_w/\mathfrak{R}$ while at the same time substituting for r^* from the Kelvin equation. Thus we obtain

$$I = \frac{p^2}{k^2 T^2}\left(\frac{2\sigma\mathfrak{m}}{\pi}\right)^{1/2}\frac{1}{\rho_L} \exp \frac{-16\pi\sigma^3}{3Z_L^3 k^3 T^3 \ln^2 S} \qquad (3.23)$$

or, with $S^2 = p^2/p_0^2$ inserted,

$$I = S^2 \frac{p_0^2}{k^2 T^2}\left(\frac{2\sigma\mathfrak{m}}{\pi}\right)^{1/2}\frac{1}{\rho_L} \exp \frac{-16\pi\sigma^3}{3Z_L^3 k^3 T^3 \ln^2 S}, \qquad (3.24)$$

which shows that the variables are the saturation ratio S and the temperature, or functions thereof. Alternatively the expression may be written logarithmically as

$$\ln I = \tfrac{1}{2}\ln\frac{2\sigma\mathfrak{m}}{\pi} - \ln \rho_L + 2\ln\frac{p_0}{kT} + 2\ln S - \frac{16\pi\sigma^3}{3Z_L^3 k^3 T^3 \ln^2 S}. \qquad (3.25)$$

For purposes of comparison with experiment it is useful to express this last equation as

$$\ln I = \ln A - B\frac{(\sigma/T)^3}{\ln^2 S}, \qquad (3.26)$$

[3] Note that N_1 has the same meaning as the symbol Z or, more specifically, Z_v (for water vapor).

which can also be written

$$B \ln \frac{I}{A} = \frac{T^3}{\sigma^3} \ln^2 S . \qquad (3.27)$$

It is seen that the quantities for which B has been substituted are essentially constant. The terms contained in A vary negligibly over a range of supersaturations up to 10, except S itself. In cloud-chamber experiments one can relate S^*, the critical saturation ratio, to the threshold of appearance of a cloud and, using the relation

$$[B \ln (I/A)]^{1/2} = (T/\sigma)^{3/2} \ln S^* = \text{const}, \qquad (3.28)$$

have a constant that can be used as an experimental verification of the theory. The constant should be the same at all threshold values of S^*. It is obvious that S^*, and therefore I and N^*, will be quite sensitive to temperature. This aspect is reconsidered in the next section.

Correction for Surface Tension

Tolman (1949) deduced from quasithermodynamic arguments that instead of the bulk surface tension a droplet surface tension defined by

$$\sigma_r = \frac{\sigma}{1 + (2\delta/r)} \qquad (3.29)$$

should be used. Here σ is the surface tension of a plane surface as used on the preceding pages and δ is the thickness of the superficial density layer as defined by Gibbs (1877). Byers and Chary (1963) demonstrated that the embryo-growth equations using this value of the surface tension produce better results. With minor terms neglected, the algebra involved in using σ_r produces a factor $1 + f^2(\ln S)$ in the last term of (3.25), where f is a function of the critical radius and the modified surface tension. The simplified expression (3.28) then becomes

$$\left(\frac{T}{\sigma}\right)^{3/2} \frac{\ln S^*}{1 + f^2(\ln S^*)^{1/2}} = \text{const} . \qquad (3.30)$$

Values of I and N^* expressed in common logarithms from the corrected equation for I at various values of T and S are given in Table 3.2. The strong dependence on temperature is apparent.

In the application of equation (3.30) to cloud-chamber measurements one is interested in the threshold of appearance of cloud which, for the same observer under given optical conditions and at the same temperature, should be fairly constant. An N^* somewhere between 1 and 1000 cm^{-3} might be the cloud threshold, depending on the crudeness of the method. As indicated in the data of Table 3.2 these values of N^* would correspond to a saturation ratio, S^*, roughly between 5 and 6. One should not expect disagreements in S^* of more than a few tenths among different observers working with the same temperatures.

47

Farley (1952) used the measurements of Powell (1928) to compute the "constant" as in equation (3.30), but for the uncorrected case without the $(1 + f^2 \ln S)^{1/2}$ in the denominator. His "constants" using σ without the correction and those with the correction factor are shown in Table 3.3. The fact

TABLE 3.2

VALUES OF CORRECTED $\text{LOG}_{10}I$ AND $\text{LOG}_{10}N^*$ FOR VARIOUS VALUES
OF SATURATION RATIO AND TEMPERATURE

TEMPERATURE °K		SATURATION RATIO (S)				
		2	4	6	8	10
253.2.....	$\log_{10}I$	− 98.7	− 6.6	7.5	12.9	15.8
	$\log_{10}N^*$	−239.8	−28.5	3.7	15.8	22.2
263.2.....	$\log_{10}I$	− 85.0	− 0.7	11.4	15.2	18.2
	$\log_{10}N^*$	−208.9	−15.6	11.8	21.8	27.0
273.2.....	$\log_{10}I$	− 71.3	4.6	14.7	18.3	20.3
	$\log_{10}N^*$	−178.2	− 4.1	18.7	26.7	31.1
283.2.....	$\log_{10}I$	− 55.7	8.7	17.2	20.2	21.9
	$\log_{10}N^*$	−142.7	4.7	23.8	30.5	34.2

TABLE 3.3

CRITICAL SATURATION RATIOS AT VARIOUS TEMPERATURES
OBSERVED BY POWELL, VALUES OF "CONSTANT" COM-
PUTED BY FARLEY, AND THE "CONSTANT" WITH THE COR-
RECTION FACTOR $(1 + f^2 \ln S^*)^{-1/2}$

Temp. after Expansion, °C	S^*	$(T/\sigma)^{3/2} \ln S^*$	$\dfrac{(T/\sigma)^{3/2} \ln S^*}{(1+f^2 \ln S^*)^{1/2}}$
47.0........	2.87	10.7	7.0
19.1........	3.74	10.6	6.8
3.2........	5.07	11.2	7.1
−16.4........	7.80	12.2	7.4
−26.4........	8.95	12.0	7.2
Mean.....	....	11.3±8.0%	7.1±4.2%

that the correction factor produces a more nearly constant quantity suggests that the Tolman correction for the surface tension is reasonably valid.

Meaning of the Theory

In order to summarize the findings concerning nucleation it is well to examine again Figures 2.3, 2.4, and 3.1. First, it is seen that in order for homogeneous nucleation to occur it is necessary that the free-energy barrier illustrated in

Figure 2.3 be surmounted. The peak of this barrier is at the Kelvin radius r^* or, if one prefers, at g^*. Figure 2.3 may be constructed in a three-dimensional model with axes ΔG, r, and S. On the r, S-plane the peaks of the barriers may be projected to form the Kelvin curve represented by the dashed line in Figure 2.4. In this figure, one may imagine that the process of homogeneous nucleation requires that an embryo of a size of a few molecules, far to the left of the border of the graph, has to grow to the critical radius represented by the Kelvin curve.

The lower the supersaturation the more growth is necessary to reach the critical radius. Figure 3.1 shows that the probability of getting embryos to this size is very low at the lower supersaturations. The equilibrium Boltzmann distribution has to be broken by a non-equilibrium "current" which, if the supersaturations are sufficiently high, can loosen the restraints. Powell's data in Table 3.3 show that saturation ratios of 3 to 9, depending on temperature, are necessary to produce noticeable effects at tropospheric temperatures. Figure 2.4 shows that when soluble nuclei are present, the situation is very different.

Critical Radius

For pure water droplets the critical radius at a given supersaturation is the single-valued equilibrium radius of the Kelvin equation. For solution droplets containing any given mass of solute, there are two equilibrium radii for a given supersaturation, one stable and the other in a labile equilibrium which, with an infinitesimal increase in the surrounding vapor pressure, will become unstable. By our previous definition of critical radius, only that on the unstable side is critical. The conditions just described are apparent from Figure 2.4.

In the homogeneous case, regardless of the degree of supersaturation, the critical radius is reached only through a chain of bombardments of vapor molecules on fortuitiously selected embryos. In the solution case there is a situation in which spontaneous growth ensues without such a chain of events. That is the situation represented by a supersaturation greater than is required for equilibrium at the peak of a curve in the r,S-plane as represented in Figure 2.4. If the peak value of the saturation ratio is exceeded by an infinitesimal amount, the solution droplet will gain water at an accelerating rate, since with this amount of supersaturation an increasingly large vapor-pressure gradient between the vapor and the droplet is produced. In this way a droplet, once over the peak, can grow to cloud sizes (r of 1 to 10 μ) in seconds. In clouds the droplet eventually reaches a limit due to the sharing of the vapor with other droplets and the consequent reduction of the supersaturation.

On the stable side of the curve, the droplet can only adjust to new equilibria as the saturation ratio changes, growing to a larger size as the humidity goes up or evaporating to a smaller size as the humidity goes down. The lower limit of the size may be considered as that corresponding to a saturated aqueous solution, reached at a saturation ratio of 0.7 in the case of a nucleus of 5×10^{-15}

49

g NaCl as represented in the inset of Figure 2.4.[4] With slowly increasing relative humidity in the natural atmosphere, such as in radiational cooling, one can note the increasing size and number of activated nuclei as evidenced by decreasing visibility. A fog in the presence of abundant nuclei is preceded and followed by haze.

In the homogeneous case the maximum elevation of the free energy, ΔG^*, for a given supersaturation was obtained by differentiating with respect to r. A simple analytical treatment of this kind is not practical for solution droplets because the molality f varies with r, and with the molality the density, the surface tension, and the electrolytic coefficient vary. By definition,

$$f = \frac{M_p \times 1000}{m_p M_w} = \frac{M_p \times 1000}{m_p(\frac{4}{3}\pi r^3 \rho_{L}' - M_p)}, \qquad (3.31)$$

where M_p and M_w are the masses of the solute and the solvent, respectively, and m_p is the molecular weight of the solute. Since M_p is constant on any given curve of Figure 2.4, and since r increases by a factor of nearly 10 between the stable and unstable sides of the curve, it is obvious that f is very sensitive to radius on the stable side but becomes very small and less sensitive to radius on the unstable side. At cloud sizes the droplets behave as pure water and even the curvature effect becomes negligible.

It should be pointed out that, at supersaturations less than the peak value of the equilibrium curve, one could follow the same processes described for homogeneous nucleation and reach the critical radius. Whereas in the case of homogeneous nucleation the process would have to start with one molecule, in this case it would start with the droplet radius on the stable side of the curve. This way of achieving the critical radius may come into prominence when the soluble particles are initially quite small, and it is obvious that energetically this is easier to accomplish here than in the homogeneous case, but when there are larger particles present which have critical radii at infinitesimally small supersaturations the peaks of the curves which one might call the *critical radii for spontaneous growth* are of overriding importance.

In chapter 5 the growth of droplets beyond the critical radius in a vapor diffusive field will be discussed.

Nucleation by Insoluble Particles

Two opposite cases of liquid-droplet nucleation have been treated: pure homogeneous nucleation with its high free-energy barrier, and nucleation by

[4] Junge (1958) finds that the salt goes into solution with increasing humidity somewhat earlier than calculated and crystallizes somewhat later with decreasing humidity. Presumably the surface is readily available for adsorption, but the molecules must move out from the interior before evaporating.

soluble particles, such as hygroscopic salts, which can occur with little or no supersaturation. Intermediate between these two opposites are insoluble particles which can nucleate at moderate supersaturations depending on their size and the extent to which their surfaces are hydrophobic or hydrophilic, "non-wettable" or "wettable." Ions form another class.

A logical way to understand the nucleation process on insoluble particles is to consider first the effects as they occur on a flat, insoluble surface.

For liquid embryos at a surface, the computation of nucleation rate requires knowledge concerning two related geometric features—the radius of curvature and the extent of spreading of the liquid over the surface. A small droplet forms a minute dome of liquid on the substrate, as shown in Figure 3.2, where a form

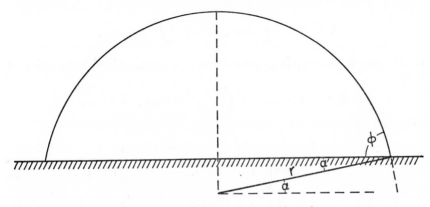

FIG. 3.2.—Water embryo on an insoluble surface

somewhat less than half a sphere is represented. It can be described by two parameters, the radius of curvature r and the angle of contact φ, as shown. The contact angle can range from 0° to 180°, the former representing a completely "wettable" surface with r infinite and the latter a completely "non-wettable" surface with the droplet standing as if it were a rigid sphere.

The following dimensions of the embryo are of interest: (1) the area on the substrate which it covers, (2) the area of the liquid-vapor interfacial surface, and (3) the volume. The distance r' of a point on the sphere from the vertical axis is $r' = r \cos \alpha = r \sin \varphi$, so the area of the substrate covered by the droplet is

$$A = \pi r^2 \sin^2 \varphi = \pi r^2 (1 - \cos^2 \varphi) . \tag{3.32}$$

The differential area of a belt around the sphere in a horizontal direction is the differential width of the belt, $r \, d\alpha$, times its circumference $2\pi r' = 2\pi r \cos \alpha$ or

$$dA_{belt} = 2\pi r^2 \cos \alpha \, d\alpha .$$

51

In terms of φ, since $a = \frac{1}{2}\pi - \varphi$ and $da = -d\varphi$, this element of area becomes $2\pi r^2 \sin \varphi \, (-d\varphi)$. The area of the portion of exposed sphere is, then,

$$A = 2\pi r^2 \int_a^{\pi/2} \cos a \, da = 2\pi r^2 \int_\varphi^0 -\sin \varphi d\varphi = 2\pi r^2 (1 - \cos \varphi). \quad (3.33)$$

In obtaining the liquid volume, distances on the vertical axis are designated as

$$y = r \sin a = r \cos \varphi ,$$

and slices of infinitesimal thickness $dy = r \cos a \, da = -r \sin \varphi \, d\varphi$ and area $\pi(r')^2 = \pi r^2 \sin^2 \varphi$ are summed by integration of

$$dV = \pi(r')^2 dy = -\pi r^3 \sin^3 \varphi \, d\varphi$$

from the surface of contact to the top of the dome; that is,

$$V = -\pi r^3 \int_\varphi^0 \sin^3 \varphi d\varphi = \pi r^3 \int_0^\varphi \sin^3 \varphi d\varphi .$$

The integral of $\sin^3 \varphi \, d\varphi$ is $\frac{1}{3} \cos^3 \varphi - \cos \varphi + $ const, and, with the indicated limits,

$$V = \pi r^3(\tfrac{1}{3} \cos^3 \varphi - \cos \varphi + \tfrac{2}{3}) = \tfrac{1}{3}(\pi r^3)(\cos^3 \varphi - 3 \cos \varphi + 2) ,$$

which factors into

$$V = \tfrac{1}{3}(\pi r^3)(2 + \cos \varphi)(1 - \cos \varphi)^2 . \quad (3.34)$$

Thus, it is seen that all quantities are functions of the radius of curvature and the cosine of the contact angle.

It is hardly necessary to give a rigorous proof of the application of the Kelvin equation to this situation, although such a proof has been outlined by Fletcher (1962a). The Kelvin equation gives the saturation ratio as a function of the curvature, $1/r$. The total volume and the area exposed to the vapor do not affect the vapor tension, which is defined for unit area, at any given curvature. Therefore the radius of curvature of the embryo determines its equilibrium vapor tension through the use of the Kelvin equation. A given number of molecules or a given volume of the liquid has a lower equilibrium vapor tension when formed into less than a sphere than when in a complete sphere because the radius of curvature is greater. The free-energy barrier is lower because the elevation of free energy is proportional to the volume as well as to the excess vapor tension (supersaturation). For a given radius of curvature r^*, the free-energy peak is lower by a factor consisting of the ratio of the volume of the embryo to that of a sphere of said radius, or, from equation (3.2),

$$\Delta G^* = \frac{16\sigma^3}{3(Z_L kT \ln S)^2} \frac{V}{V_s} = \frac{4\sigma^3}{3(Z_L kT \ln S)^2} (2 + \cos \varphi)(1 - \cos \varphi)^2, \quad (3.35)$$

where V_s is the volume of a sphere of radius r^*.

In the calculation of the nucleation rate the same distribution as used in the homogeneous case is applied:

$$N(r^*) = N_1' \exp(-\Delta G^*/kT), \tag{3.36}$$

where the N's now signify numbers per unit area of surface. The approximation can be made for the area of nucleation on the substrate (eq. [3.20])

$$A = \pi r^{*2}(1 - \cos^2 \varphi) \approx \pi r^{*2}, \tag{3.37}$$

which is exact for a hemispherical embryo and reasonable for contact angles between $\pm 60°$ and $90°$. The rate of impact of molecules per unit area, as obtained from (3.17), is $\mathfrak{N}p(2\,m_wRT)^{-1/2}$ or $p(2\mathfrak{m}_wkT)^{-1/2}$, where $\mathfrak{m}_w$ is the mass of a molecule of water. The nucleating rate per unit area of substrate becomes

$$J_A = \frac{\pi r^{*2}p}{(2\mathfrak{m}_wkT)^{1/2}} N_1' \exp \frac{-4\sigma^3}{3Z_L^2k^3T^3 \ln^2 S}(2 + \cos \varphi)(1 - \cos \varphi)^2. \tag{3.38}$$

This expression determines the rate with respect to direct condensation from the vapor. Pound et al. (1954) deduced that molecules are adsorbed first on the surface, then diffuse along the surface to an embryo site. The nucleating rate under these conditions is hundreds of times faster than that obtained from vapor kinetics alone. Fletcher (1962b), considering that the adsorption constitutes an appreciable fraction of a monolayer, finds the kinetic coefficient to be of the order of 10^{24} to 10^{27} cm^{-2} sec^{-1} at the required supersaturations and normal temperatures. Under these conditions the critical saturation ratios as a function of contact angle can be determined for a given nucleation rate. For a rate of 1 cm^{-2} sec^{-1}, contact angles of $45°$ to $90°$ correspond to saturation ratios lying in the approximate range of 1.5 to 3. At $\varphi = 0$ the surface is wetted and the equilibrium is that of a plane water surface ($S = 1$), while at $\varphi = 180$ the nucleation is theoretically homogeneous.

In treating an insoluble *particle* as a nucleus, one assumes that it is a sphere of radius r_0. A factor appears in the expression for the free-energy difference ΔG^* which is considerably more complicated than that for a flat surface. The expression, without elaboration of the modifying factor, is

$$\Delta G^* = \frac{16\pi\sigma^3}{3(Z_LkT \ln S)^2} f\left(\cos \varphi, \frac{r_0}{r^*}\right). \tag{3.39}$$

Fletcher (1958a) developed the $f(\cos \varphi, r_0/r^*)$ factor. With $r_0/r \equiv x$ and $\cos \varphi \equiv y$, his result may be written as

$$f\left(\frac{r_0}{r^*}, \cos \varphi\right) = f(x, y) = 1 + \left(\frac{1 - xy}{\lambda}\right)^3 \tag{3.40}$$

$$+ x^3\left[2 - 3\left(\frac{x - y}{\lambda}\right) + \left(\frac{x - y}{\lambda}\right)^3\right] + 3x^2y\left(\frac{x - y}{\lambda} - 1\right),$$

where $\lambda = (1 + x^2 - 2xy)^{1/2}$. Instead of the nucleating rate per unit area as in (3.38), we may write the nucleating rate for the particle by inserting its area. We have, then,

$$J \approx \frac{4\pi^2 r^2 r^{*2}}{(2\pi m kT)^{1/2}} N_1' \exp\left(\frac{-16\pi\sigma^3}{3Z_L^2 k^3 T^3 \ln^2 S}\right) f(x, y). \quad (3.41)$$

Fletcher (1958a) determined that the pre-exponential factor has the approximate value $10^{25} \times 4\pi r^2$, except when $r_0 < r^*$, in which case it is appropriate to replace r_0 with r^*. He computed the critical saturation ratio for a J of one particle nucleating a droplet per second as a function of particle radii and for several values of the cosine of the contact angle. The result is shown in Figure

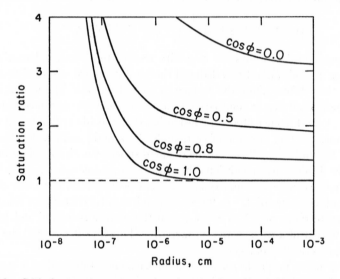

FIG. 3.3.—Critical saturation ratio as a function of radius of insoluble particle for various values of contact angle. (After Fletcher, 1958.)

3.3. The obvious conclusion is that in order for a particle to be active as a condensation nucleus it must be of reasonably large size, but must also permit water to form on it with low contact angle.

An expression developed by Krastanow (1948) contains ΔG^* as given in (3.20), with a term subtracted for the effect of the nucleus, as follows:

$$\Delta G^* = \tfrac{4}{3}\pi r^{*3}\sigma - \tfrac{4}{3}\pi r^2\sigma[3 \cos \varphi - 2(r/r^*)]. \quad (3.42)$$

Miloshev and Krastanow (1963) find this expression to be reasonably accurate for small nuclei and for contact angles close to complete wetting.

Nucleation by Ions

In his classical cloud-chamber experiments in 1897 and 1899 C. T. R. Wilson found that in absolutely clean air condensation occurred at supersaturations less than those required for completely homogeneous nucleation. He correctly explained this effect as being due to the presence of ions in the atmospheric gases and vapors. Over the years this effect has been examined in detail.

As a charge q moves from a central point to the surface of a sphere of radius r, work is done through an electrical-potential difference $\Delta\phi = q/\kappa r$ (electrostatic units), where κ is the dielectric constant of the medium. The work differential is $dW = \Delta\phi \; dq = (\kappa r)^{-1}q \; dq$ which, integrated, gives the work—equivalent to the increase in potential energy—as $W = q^2/2\kappa r$. Tohmfor and Volmer (1938) pointed out that when 10 to 100 molecules of water go onto an ion, the radii of the droplet and the ion both need to be considered, so that the potential energy becomes proportional to $(q^2/2r_0) - (q^2/2r)$, where r_0 is the radius of the ion. Furthermore, the charges are carried in different media with dielectric constants κ_0 for air (vapor) and κ for water. The potential energy which contributes a change in surface free energy is, then,

$$\frac{q^2}{2\,\kappa_0 r_0} - \frac{q^2}{2\,\kappa r} = \frac{q^2}{2}\left(\frac{1}{r} - \frac{1}{r_0}\right)\left(\frac{1}{\kappa_0} - \frac{1}{\kappa}\right). \qquad (3.43)$$

The total elevation of the free energy above that for a standard water surface becomes

$$\Delta G = -\tfrac{4}{3}\pi r^3 Z_L kT \ln\frac{p}{p_0} + 4\pi r^2\sigma + \frac{q^2}{2}\left(\frac{1}{\kappa_0} - \frac{1}{\kappa}\right)\left(\frac{1}{r} - \frac{1}{r_0}\right). \qquad (3.44)$$

Equilibrium with the vapor is, as before, at the maximum given from the derivative with respect to radius equated to zero, a condition which is satisfied by

$$\ln\frac{p}{p_0} = \frac{2\sigma}{Z_L kTr} - \frac{q^2}{8\pi Z_L kTr^4}\left(1 - \frac{1}{\kappa}\right), \qquad (3.45)$$

where the dielectric constant κ_0 is taken as one.

The dielectric constant of bulk water is between about 75 and 100 at atmospheric temperatures, so the $1/\kappa$ factor often is ignored. However, in the collection of some tens of molecules forming the nucleating mass in a strong field around the ion, κ may approach values of less than 10 as suggested by refraction and other optical effects. Tohmfor and Volmer proposed a value of 1.85, which produces agreement with experimentally determined rates of formation of droplets in cloud chambers.

Curves of the equilibrium saturation ratio over ionized droplets are shown in Figure 3.4 for different assumed values of the dielectric constant of the water embryos. While the curves look similar to those for hygroscopic nuclei, it should be noted that the scale of representation is vastly different from that of Figure

2.4, being much higher in supersaturation and in a range of droplets smaller by 10^{-3} to 10^{-4}.

Droplets condensing on ions can be carried to the critical radius at saturation ratios less than those at the peaks of the curves in Figure 3.4 by the same process that prevails in homogeneous nucleation, except that the growth starts at the radius on the stable side of the curve. Equation (3.22) applied in this case is

$$I = f(p, T) \, N \, \exp\left(- \, \Delta G^*/kT\right), \qquad (3.46)$$

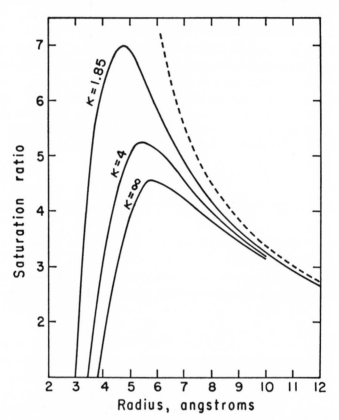

Fig. 3.4.—Equilibrium curves for spherical droplet embryos bearing unit electronic charge for various values of the dielectric constant, κ, at $0°$ C. Dashed line is Kelvin curve for uncharged embryos.

where N is the number of ions per cm^3. Tohmfor and Volmer (1938) ascribed a value of 10^7 to $f(p, T)$ at $265°$ K, developed in a manner similar to that used earlier in this chapter. To make these values fit the observation of an S^* of about 4 at this temperature, they introduced a κ of 1.85 into ΔG^*.

The theoretical treatments fail to account for a regularly observed difference in S^* for negative as against positive ions. Wilson noted this effect in his 1899 experiments, finding that at about $-6°$ C the S^* was only about 4 for the negative ions while, for positive ions at about the same temperature, it was about 6. The difference is usually explained qualitatively on the basis of evidence that the dipole of the water molecule has its negative oxygen end oriented outward at the surface, and in the first two or three molecular layers. With an extra negative ion the energy of these dipoles is reduced and, consequently, there is a lowering of the surface free energy. Positive ions have the opposite effect.

Homogeneous Nucleation of Ice

While substantial cooling below the dew point does not occur in the natural atmosphere, marked undercooling of water droplets below $0°$ C is the rule rather than the exception. One infers from this circumstance that there are nuclei which are active in heterogeneous nucleation of the liquid phase but have much less particle activity for heterogeneous nucleation of the ice phase. A temperature of approximately $-40°$ C seems to be fixed as the temperature of freezing of the purest of small water droplets. The soluble salts that are active in condensation serve to lower the freezing point below this figure, indicating that the nuclei for freezing are not the same as for condensation. Nuclei which cause the formation of ice clouds at temperatures of $-10°$ to $-20°$ are often available in the atmosphere, but they are not the same particles that are active in condensation.

The theory of homogeneous nucleation of ice should account for the occurrence of the pure-water phenomenon at $-40°$, sometimes referred to as the Schaefer point from the work of V. J. Schaefer (1949) who brought it into prominence through demonstrations in a laboratory cold box in the 1940's, although Cwilong (1947) appears to have been first to publish experimental results.

One can conceive of the homogeneous formation of embryo ice crystals directly from the vapor in clouds (Cwilong, 1947; Bradley, 1951; Schaefer, 1952) or consider that they form by freezing in or on undercooled liquid water droplets (Fisher *et al.*, 1949; Lafargue, 1950; Mason, 1952, 1957). From thermodynamic considerations, Krastanow (1941) showed that water vapor will condense to undercooled liquid more easily from an energetic point of view than it will go directly to the solid state. He computed that direct vapor $\rightarrow$ ice crystallization would not occur until a temperature of $-65°$ C or lower was reached. Data from experiments of Sander and Damköhler (1943) and Pound *et al.* (1951) have shown a discontinuity at $-62°$ C and $-63°$ C, respectively, which Fletcher (1962*b*) interprets as probably marking the "sublimation threshold." Since at temperatures above $-40°$ it is not possible to create a cloud, at least in the laboratory, without many droplets being present, it is virtually impossible to

57

detect direct deposition even if it should occur. The theoretical treatment which follows is based on the assumption that the ice forms within the liquid.

The theory has evolved mainly from analyses by Becker and Döring (1935), Frenkel (1946), Turnbull and Fisher (1949), Bradley (1951), McDonald (1953), and Fletcher (1962a). The expression for the steady current operates directly through a Boltzmann type of distribution, having elements of similarity with equation (3.4) but lacking the same embarrassing situation at the large sizes.

The rate of acquisition of molecules by embryos cannot be stated in terms of kinetic theory as in the vapor case, because in the liquid the molecules are limited in their motions by binding forces. A term expressing this rate has been determined by Turnbull and Fisher to be approximately

$$(kT/h) \exp\left(-\Delta G'/kT\right), \tag{3.47}$$

where h is Planck's constant[5] and $\Delta G'$ is the activation energy of self-diffusion of the liquid across the liquid-ice boundary. In order to diffuse through the liquid across the interface a molecule must break one or more of the tetrahedral bonds with which it is interlocked with other molecules. The energy, $\Delta G'$, required to reach this freer state, called the activation energy for self-diffusion, is obtained from random thermal-vibrational processes in the liquid having a probability given by $\exp(-\Delta G'/kT)$.

As deduced by Turnbull and Fisher, the rate of formation of ice particles per cm³ per second is

$$J \approx (Z_L kT/h) \exp\left[-(\Delta G' + \Delta G_e^*)/kT\right], \tag{3.48}$$

where Z_L is the number of molecules per unit volume in the liquid phase and ΔG_e^* is the free energy of formation of an embryonic ice nucleus of equilibrium size. For this free-energy increment we have an expression similar to (3.1). However, the embryo is likely not to be spherical, so we introduce geometric factors a and b such that the surface area of the embryo is given by $4\pi a r^2$ and the volume by $4\pi b r^3/3$. The expression is a slightly modified (3.1) or

$$\Delta G_e = 4\pi a r^2 \sigma_e - \tfrac{4}{3}\pi b r^3 Z_e kT \ln\left(p_L/p_e\right), \tag{3.49}$$

where σ_e is the specific surface free energy of the ice-liquid interface, Z_e is the number of molecules of ice per cm³, and p_L and p_e are the saturation vapor pressures over liquid water and over ice, respectively, at temperature T.

To find the equilibrium ΔG_e^*, we determine the maximum by taking the derivative as in (3.2) and obtain

$$8\pi a r \sigma_e = 4\pi b r^2 Z_e kT \ln\left(p_L/p_e\right) \tag{3.50}$$

for the maximum. Then, as in the derivation of (3.20), we find

$$\Delta G_e^* = 4\pi a r^{*2} \sigma_e/3. \tag{3.51}$$

[5] Note that with h inserted the rate is in terms of units of free energy per unit time.

One may simplify the expression by substituting a geometrical factor $a' = 4\pi a$. The r^* may be a characteristic length for the crystal. The equilibrium r^*, as obtained from (2.42), is

$$r^* = \frac{2 a \sigma_e}{b Z_e k T \ln (p_L/p_e)}. \tag{3.52}$$

Instead of referring the critical size to the ratio of the saturation vapor pressures, it is more convenient to express it in terms of the undercooling below the melting temperature. Frenkel (1946) used the expression

$$r^* = \frac{2 \sigma_e T_0}{\rho_e L_f (T_0 - T)}, \tag{3.53}$$

where T_0 is the nominal freezing point of $273°$ K, ρ_e is the density of the ice, and L_f is the latent heat of fusion. Since L_f is a function of T, Dufour and Defay (1963) introduced $\bar{L}_f$, an arithmetic mean of $L_f(T_0)$ and $L_f(T)$. McDonald (1964) suggests a method of obtaining this expression—and corrects an erroneous result of his (McDonald, 1953)—by integrating the Clapeyron-Clausius equation along the two paths of vapor to ice and vapor to liquid at subfreezing temperatures.

The integral is performed from p_L, T to the triple point p_t, T_0 and from p_e, T to the same point. Expressing the Clapeyron-Clausius equation for one gram as

$$\left(\frac{dp}{p}\right)_{L, v} = \frac{m_w L_v}{R} \frac{dT}{T^2} \quad \text{and} \quad \left(\frac{dp}{p}\right)_{e, v} = \frac{m_w L_s}{R} \frac{dT}{T^2}, \tag{3.54}$$

one forms the integrals

$$\int_{p_L}^{p_t} \frac{dp}{p} = \frac{m_w \bar{L}_v}{R} \int_{T}^{T_0} \frac{dT}{T^2}, \tag{3.55}$$

$$\int_{p_e}^{p_t} \frac{dp}{p} = \frac{m_w \bar{L}_s}{R} \int_{T}^{T_0} \frac{dT}{T^2}, \tag{3.56}$$

where $\bar{L}_s$ and $\bar{L}_v$ are the mean values that will make the integrals correct. The results are

$$\ln \frac{p_t}{p_L} = \frac{m_w \bar{L}_v}{R} \left(\frac{1}{T} - \frac{1}{T_0}\right) = \frac{m_w \bar{L}_v}{R} \frac{T_0 - T}{T T_0}, \tag{3.57a}$$

$$\ln \frac{p_t}{p_e} = \frac{m_w \bar{L}_s}{R} \left(\frac{1}{T} - \frac{1}{T_0}\right) = \frac{m_w \bar{L}_s}{R} \frac{T_0 - T}{T T_0}, \tag{3.57b}$$

from which p_t can be eliminated to produce

$$\ln \frac{p_L}{p_e} = \frac{m_w (\bar{L}_s - \bar{L}_v)}{R} \cdot \frac{T_0 - T}{T T_0} = \frac{m_w \bar{L}_f}{R} \cdot \frac{T_0 - T}{T T_0}, \tag{3.58}$$

where use is made of the empirical relation $L_s = L_f + L_v$, mentioned in chapter 2. This expression may be substituted in (3.52), recognizing that $Z_e k = \rho_e R/m_w$, to obtain, for a spherical geometry ($b = 1$),

$$r^* = \frac{2\sigma_e T_0}{\rho_e \bar{L}_f (T_0 - T)}, \tag{3.59}$$

which is Frenkel's equation (3.53).

Putting the value of r^* into (3.51) and then substituting this expression into (3.48), we obtain an expanded equation for the rate

$$J \approx \frac{Z_L k T}{h} \exp\left\{-\left[\frac{\Delta G'}{kT} + \frac{16\pi a \sigma_e^3 T_0^2}{3kT \rho_e^2 \bar{L}_f^2 (T_0 - T)^2}\right]\right\}. \tag{3.60}$$

One notes that the critical dependence of ΔG_e^* on the temperature as shown in (3.51) through use of (3.59) causes J to increase with the undercooling despite the opposing effect of the kT in both the exponent and the pre-exponential factor of (3.60). It is apparent that the less the undercooling the larger the embryo has to be for equilibrium, and the larger must be the negative $\Delta G_e^*/kT$ in the exponent.

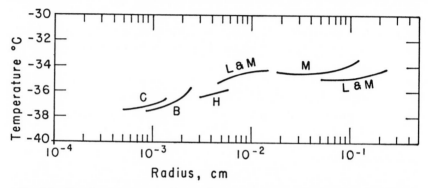

FIG. 3.5.—Freezing temperature of water drops as a function of their size as determined by various investigators: *B*, Bigg; *C*, Carte; *H*, Hoffer; *L&M*, Langham and Mason; *M*, Mossop.

There are a number of difficulties in computing values of J. The specific free energy σ_e of the liquid-ice interface can be determined only to within a factor of about two, corresponding to an even greater uncertainty in the degree of undercooling. Another difficulty concerns the growth habit or geometry of the ice embryo, that is, the factor a in (3.65). Furthermore, the activation energy $\Delta G'$ for self-diffusion in the liquid has not been determined at low temperatures.

Pure bulk water cannot be undercooled significantly; but, when it is divided into small drops, its freezing temperature becomes much lower. It is not surprising that there should be a volume dependence of the freezing temperature. Even in the purest water some foreign particles might exist to help the nucleation, but the smaller the drop the less the chance that it might contain such a particle. The process is of a statistical nature—the smaller the drop the less the statistical probability that it will freeze at a given temperature, and the lower the temperature the greater the probability that a drop of a given size will be frozen.

60

Several investigators (Bigg, 1953; Mossop, 1955; Carte, 1956; Langham and Mason, 1958; Hoffer, 1961) have taken water of the most extreme purity, made it into drops, and studied what appeared to be homogeneous nucleation. The best data were obtained by floating the drops in silicone oil or similar liquid to eliminate effects of solid surfaces. The results of some of these findings, which show a decreasing freezing temperature with decreasing size, are given in Figure 3.5. In Hoffer's study soluble salts, such as form condensation nuclei, were found to lower the freezing temperature when placed in pure water. Otherwise it was quite evident that the purer the water the lower the freezing temperature.

Heterogeneous Nucleation of Ice Crystals

Nucleation of the ice phase has been treated in the preceding section as occurring through the liquid phase in droplets. In the case of heterogeneous nucleation the process operates after a thin film of water has been adsorbed on the surface of the nucleus. The film freezes as a result of nucleation of ice by

TABLE 3.4

COMPARISON OF ICE-NUCLEATING CRYSTALS WITH ICE CRYSTALS*

CRYSTAL SUBSTANCE	LATTICE CONSTANTS, ANGSTROMS		BASAL MISFIT %	PRISM MISFIT %	REPORTED NUCLEATING THRESHOLD, °C
	a	b			
Ice.........	4.52	7.36	....	....	....
AgI........	4.58	7.49	1.4	1.6	−4
PbI$_2$.......	4.54	6.86	0.5	3.6	−6

* All crystals have a hexagonal form.

the particle surface, but the film is so thin—*ca.* 100 A—that no droplets are detected and an ice crystal forms as if from the vapor.

The ice crystals form selectively on suitable surfaces where a matching of lattice structure and molecular distances, called *epitaxy*, is present. The ice crystals grow with their axes parallel to corresponding orientations in the lattice of the substrate. If there is a small misfit between the molecules of the ice embryo and the substrate at the interface, a dislocation of molecules may occur. It is to be expected that an epitaxy which gives a minimum of such dislocations would be favored because it would produce the lowest interfacial free energy.

In Table 3.4 are listed, in comparison with ice, some crystals which are known to be active nucleating agents for ice crystals. Silver iodide, for example, is a substance commonly used in artificial nucleation of clouds in the laboratory and in the open atmosphere. The basal hexagonal or *a*-plane and the prism or *c*-axis are compared between the nucleating crystal and the ice crystal. The substances are essentially insoluble in water.

Theory of Heterogeneous Ice Nucleation

It is possible to formulate a theory of ice nucleation on particles along the lines used in developing the theory of nucleation of liquid condensation on insoluble particles. Fletcher (1958*a*, 1963) has carried the problem through to reasonable solutions, using as a starting point the general expression for the nucleating rate

$$J = K \exp\left(-\Delta G^*/kT\right). \tag{3.61}$$

From the usual kinetic arguments he finds that for unit area $K_A \approx 25 \text{ cm}^{-2} \text{ sec}^{-1}$. The crux of the problem is to obtain ΔG^*. In the liquid case this was done by considering the contact angle (wettableness) and ratio of radii. Analogous to the wettableness is the compatibility between the ice and the crystalline nucleus, while characteristic dimensions can be applied in place of radii.

The measure of compatibility should be approached from a molecular point of view, but it is found that the bulk property of surface tension will be descriptive of the process and will lead to a useful result even when used crudely. The compatibility factor may be expressed as

$$y = \frac{\sigma_{LP} - \sigma_{IP}}{\sigma_{LI}}, \tag{3.62}$$

where the subscripts refer to interfaces involving liquid (L), ice (I), and particle (P). Actually it can be shown in the vapor-liquid-particle system treated before that this factor is exactly equal to the cosine of the angle of contact. For a nucleus of characteristic dimension R and an ice embryo of critical dimension r^*, let $x = R/r^*$, then

$$\Delta G^* = \Delta G_0^* f(x, y), \tag{3.63}$$

where ΔG_0^* is the critical free energy for homogeneous nucleation. As in the liquid condensation case, $f(x, y)$ is a complicated function of the geometry of the situation.

For a spherical ice embryo on a spherical particle, $f(x, y)$ is the same as in the embryo droplet case with y replacing $\cos \varphi$, except that Fletcher chooses to define it with a factor of $\frac{1}{2}$, e.g.,

$$\tfrac{1}{2} + \tfrac{1}{2}[(1 - xy)/g]^3 + \ldots,$$

etc., where, as before, $g = (1 + x^2 - 2xy)^{1/2}$. Fletcher also considers an embryo of cylindrical habit on a sphere, a cylinder on a disklike cylinder, and a cylinder on a needle-like cylinder, each with a complicated $f(x, y)$ which will not be repeated here.

For a spherical embryo, $\Delta G_0{}^*$ may be written, after the manner of (3.2), as

$$\Delta G_0{}^* = \frac{16\pi\sigma_{LI}{}^3}{3(\Delta G_v)^2},\tag{3.64}$$

where ΔG_v is the free energy utilized in the formation of unit volume of the new phase (ice). In the same context, critical radius is

$$r^* = \frac{-2\sigma_{LI}}{\Delta G_v}.\tag{3.65}$$

For an n-sided prismatic embryo, Fletcher (1960) finds

$$\Delta G_0{}^* = \frac{8\pi\sigma_{LI}{}^3\xi}{(\Delta G_v)^2},\tag{3.66}$$

where $\xi = (n/\pi)\tan(\pi/n)$.

To arrive at a general form, one may take equation (3.66) with ξ assumed to be 1 and write

$$(\Delta G_v)^2 = \frac{8\pi\sigma^3}{\Delta G_0{}^*} = \frac{8\pi\sigma^3 f(x,y)}{\Delta G^*},\tag{3.67}$$

where σ stands for σ_{LI}. The ΔG^* may be obtained from (3.61), assuming that a rate of $J \approx 1$ sec^{-1} is an appropriate threshold. The factor K_A, as stated before, may be taken as 10^{25} cm^{-2} sec^{-1}, and the area of the type of nucleus considered by Fletcher is approximated by $10R^2$ so that $K \approx 10^{26}\,R^2$. Under these assumptions

$$\Delta G^* = kT(60 + 2\ln R),\tag{3.68}$$

$$(\Delta G_v)^2 = \frac{8\pi\sigma^3 f(x,y)}{kT(60 + 2\ln R)}.\tag{3.69}$$

With x and r^* in terms as defined and as given in (3.65), and with some rearrangement,

$$\frac{(\Delta G_v)^2}{\sigma^3} = \frac{8\pi f(y, -\frac{1}{2}R^{1/2}\Delta G_v\sigma^{-3/2})}{kT(60 + 2\ln R\sigma^{1/2} - \ln\sigma)}.\tag{3.70}$$

Fletcher introduced the quantities $z = \Delta G_v\sigma^{-3/2}$ and $\zeta = R\sigma^{1/2}$, then taking $T = 300°$ K and $\ln\sigma = 5$, he found

$$z^2 \approx \frac{3.0 \times 10^{14} f(y, -\frac{1}{2}\zeta z)}{27 + \ln\zeta}.\tag{3.71}$$

By a similar process for a spherical embryo he obtained the same equation, except that the factor 3.0 is replaced by 2.0. The results were given in plots of ζ against z.

One can obtain the supersaturation or undercooling as a function of R by using the relationship

$$\Delta G_v = -Z_e kT \ln(p/p_0)\tag{3.72}$$

for deposition from the vapor, and for freezing,

$$\Delta G_v \approx - \Delta S_v \Delta T, \tag{3.73}$$

where ΔS_v is the entropy of melting per unit volume and ΔT is the undercooling.

A phenomenon well known in crystallography is the favored growth of embryos of the new phase on irregular features of the nucleating surface. Cavities or steps can produce an enhanced nucleating effect. For substances which are not ordinarily good nucleators the enhancement effect may be quite striking. Steps or similar edges are quite common in crystalline substrates. One configuration that has attracted considerable attention to some substances is a spiral-edged flat cone extending from a crystal face, which presents a continuous edge for rapid crystal overgrowth oriented around the spiral axis.

Among the better known experiments of the growth of ice crystals on crystalline substrates are those of Hallett (1961; see also Hallett and Mason, 1958; Bryant *et al.*, 1959). Ice crystals were grown on freshly cleaved surfaces of covellite (natural cupric sulphide), and since they were often very thin they gave rise to brilliant interference colors in translucent illumination. Striking color changes accompanied the thickening of the crystals permitting a determination of the rate of growth.

The first crystals to appear on the covellite invariably grew along those steps in the substrate which exceeded about 0.1 μ in height. Crystals could be found growing on flat areas of the surface when there was a large excess vapor density, substantiating the known compatibility of ice and CuS. A typical hexagonal plate of constant thickness of 0.6 μ grew in diameter at a rate of nearly 0.5 μ per second with a vapor excess of 0.39×10^{-6} g cm^{-3} at $-14°$ C.

Head (1961) studied experimentally and theoretically the effects of topography on a compatible substrate—AgI—and a poorly nucleating substrate—CdS, contact angle 90°. On the silver iodide, ice grew at ledges formed where the sloping sides of hexagonal etch pits met the base, but no lessening of the undercooling in general seemed to occur, the ice always showing an onset temperature at the customarily quoted threshold of $-4°$ C.

At the lowest temperature of his apparatus, $-25°$ C, Head found no freezing activity on the flat surfaces of cadmium sulphide. With stepped crystals, however, some nucleations occurred, particularly in the smaller steps, in the range $-15°$ C to $-20°$ C. If the faces were roughened by etching, nucleations occurred in the range $-12°$ C to $-18°$ C. Broken crystals sometimes showed nucleation at fractures, and some fragments produced by crushing showed an onset of ice embryos at $-10°$ C.

REFERENCES

BECKER, R., and DÖRING, W. (1935), *Ann. Phys.*, **24**, 719.

BIGG, E. K. (1953), *Proc. Phys. Soc., Ser. B*, **66**, 688.

BRADLEY, R. S. (1951), *Quart. Rev.*, **5**, 315.

BRYANT, G. W., HALLETT, J., and MASON, B. J. (1959), *J. Phys. Chem. Solids*, **12**, 189.

BYERS, H. R., and CHARY, S. K. (1963), *Zeitschr. angew. Math. Phys.*, **14**, 428.

CARTE, A. E. (1956), *Proc. Phys. Soc., Ser. B*, **69**, 1028.

CWILONG, B. M. (1947), *Proc. Roy. Soc., London, Ser. A*, **190**, 137.

DUFOUR, L., and DEFAY, R. (1963), *Thermodynamics of Clouds* (New York: Academic Press), p. 144.

FARLEY, F. J. M. (1952), *Proc. Roy. Soc., London, Ser. A*, **212**, 530.

FISHER, J. C., HOLLOMAN, J. H., and TURNBULL, D. (1949), *Science*, **109**, 168.

FLETCHER, N. H. (1958), *J. Chem. Phys.*, **29**, 572, and **31**, 1136.

———. (1960), *Australian J. Phys.*, **13**, 408.

———. (1962*a*), *Phil. Mag.*, **7**, 255.

———. (1962*b*), *The Physics of Rainclouds* (Cambridge: Cambridge University Press), pp. 53 and 210.

———. (1963), *J. Chem. Phys.*, **38**, 237.

FRENKEL, J. (1946), *Kinetic Theory of Liquids* (Oxford: Clarendon Press).

GIBBS, J. W. (1877), *Trans. Conn. Acad.*, **3**, 343, or *Collected Works* (New York: Longmans, Green and Co., 1928), **1**, p. 219.

HALLETT, J. (1961), *Phil. Mag.*, **6**, 1073.

HALLETT, J., and MASON, B. J. (1958), *Proc. Roy. Soc., London, Ser. A*, **247**, 440.

HEAD, R. B. (1961), *Bull. Obs. Puy de Dôme*, No. 1, p. 47.

HOFFER, T. E. (1961), *J. Met.*, **18**, 766.

JUNGE, C. E. (1958), in *Advances in Geophysics*, ed. H. LANDSBERG and J. VAN MIEGHEM (New York: Academic Press), **4**, 1.

KRASTANOW, L. (1941), *Met. Zeitschr.*, **58**, 37.

———. (1948), *Annuaire Univ. Sofia*, **44**, No. 1.

LAFARGUE, C. (1950), *C. R. Acad. Sci. Paris*, **246**, 1894.

LANGHAM, E. J., and MASON, B. J. (1958), *Proc. Roy. Soc., London, Ser. A*, **247**, 493.

MCDONALD, J. E. (1953), *J. Met.*, **10**, 416.

———. (1963), *Amer. J. Phys.*, **31**, 31.

———. (1964), *J. Atmos. Sci.*, **21**, 225.

MASON, B. J. (1952), *Quart. J. Roy. Met. Soc.*, **78**, 22.

———. (1957), *The Physics of Clouds* (New York: Oxford University Press).

MILOSHEV, G., and KRASTANOV, L. (1963), *Tellus*, **15**, 297.

Mossop, S. C. (1955), *Proc. Phys. Soc., Ser. B*, **68**, 193.

Pound, G. M., Madonna, L. A., and Sciulli, C. (1951), Carnegie Inst. Tech. Metals Res. Lab. Quart. Rep. No. 5.

Pound, G. M., Simmad, M. T., and Yang, L. (1954), *J. Chem. Phys.*, **22**, 1215.

Powell, C. F. (1928), *Proc. Roy. Soc. London, Ser. A*, **119**, 553.

Sander, A., and Damköhler, G. (1943), *Naturwissensch.*, **31**, 460.

Schaefer, V. J. (1949), *Chem. Rev.*, **44**, 291.

———. (1952), Project Cirrus, Gen. Elec. Res. Lab., Rep. No. 33.

Tohmfor, G., and Volmer, M. (1938), *Ann. Phys., Ser. 5* (Leipzig), **33**, 109.

Tolman, R. C. (1949), *J. Chem. Phys.*, **17**, 333.

Turnbull, D., and Fisher, J. C. (1949), *J. Chem. Phys.*, **17**, 71.

Volmer, M., and Weber, A. (1926), *Zeitschr. Phys. Chem.*, **119**, 277.

Wilson, C. T. R. (1897), *Phil. Trans., Ser. A*, **189**, 265.

———. (1899), *ibid.*, **193**, 289.

4

NUCLEI
IN THE ATMOSPHERE

In the perpetual interactions between the surface of our planet and the lower atmosphere there is a continuous exchange of gaseous components and of liquid and solid particulates as well. Locally these processes are strongly augmented by man's activities. From the outer atmosphere another class of particulates mainly thought of as meteoritic dust is contributed.

Depending on their physical and chemical nature, particles are suspended for a period of time varying between minutes and years. Gravitational fallout ceases to be significant for particles of a diameter less than about 3μ. At this size begins the range of the aerosols. They might remain airborne indefinitely were it not for their occasional encounter with clouds and precipitating water which washes some of them out of the atmosphere.

While the upper limit of size of aerosols is determined by gravitational removal, the lower limit is not clearly defined. The definition involves the borderline between a molecule in the free, gaseous state and an embryo particle. For substances of appreciable vapor pressure, sizes below the critical or equilibrium radius given by the Kelvin relation have a low probability of occurrence and would be only transitory because of their high volatility relative to the larger particles of the same substance.

The most common limitation of size on the lower end of the scale comes about through coagulation. In this process the smallest, most mobile members collide and stick together to form larger, less mobile and thus more stable particles.

The various processes in the lower atmosphere have the net effect of producing a concentration of aerosols in the diameter range of roughly 0.01 to 1μ of about

100 to 10,000 per cm³. Beyond this range of sizes the concentrations fall off rapidly.

From measurements at the ground, Junge (1958) has equated size and number of aerosols to their total volume in each size class and has found that the volume of aerosol per given volume of air is about the same in all size classes from about 0.1 to 10μ radius. Thus $Nr^3 = $ const and the concentration N increases by three orders of magnitude for each order of magnitude decrease in r, or $d(\log N)/d(\log r) = -3$. This distribution is illustrated in Figure 4.1 of the next section.

Ions form a special class of aerosol. They are continuously being created in the atmosphere by the cosmic radiation and over the continents from the radioactive gases emanating from the soil. Under favorable conditions there may be 100 to 1000 small ions per cm³ near sea level, increasing with height to about 10^5 or more per cm³ in the ionosphere. There is an inverse relation between the number of small ions and the number of larger aerosols at any one time and place. The common aerosols capture and largely immobilize the ions. In fact, it is found that the more numerous aerosols provide an effective means of reducing the number of small ions by this capture process. About half of the areosols in the 0.01 to 0.1μ range carry a net charge as a result of ion capture and therefore are classed as large ions. They have mobilities of less than 10^{-3} cm² volt⁻¹ sec⁻¹ contrasted with the mobilities of about 1.4 cm² volt⁻¹ sec⁻¹ for the small ions.

Small ions become important for condensation only when the air is cleaned of other aerosols. The large supersaturations required before they can serve as centers of condensation make them ineffective in the natural atmosphere. High water-vapor content and near–micron-sized aerosols usually are found together because they both come from the same source—the surface of the earth.

Nucleating Activity

The activity of a collection of particles of the same substance and of the same size in nucleating water droplets is most conveniently stated in terms of the saturation ratio at which a noticeable fog occurs when the particles are present in otherwise clean air. The fog is usually created in a chamber by cooling through expansion. For a given initial humidity the chamber must be calibrated for saturation ratio as a function of expansion ratio.

Our knowledge of atmospheric particles is largely founded on the work of John Aitken from 1884 to 1912. Through his portable expansion chamber, still known as the Aitken nuclei counter, he surveyed the particle content of the lower atmosphere. His equipment produced supersaturations at which the most numerous particles, those between 0.01 and 0.1μ radius, are active. Today these are generally spoken of as Aitken particles. They are capable of serving as condensation nuclei for clouds and fogs, but it is now believed by cloud physicists that nuclei larger than those measured in the Aitken counter are sufficiently numerous in the lower half of the troposphere to lay first claim to the available water. Hygroscopic Aitken nuclei, that is, those composed of substances having a chemical affinity for water, require supersaturations of roughly 0.5 to 2.0 per cent in order to carry water through the vapor-to-liquid transition. Large or "cloud" nuclei with radii in the next higher order, 0.1 to 3μ, require supersaturations of less than 0.5 per cent at the lower end of the size range to undersaturation for hygroscopic substances in the upper half of the range.

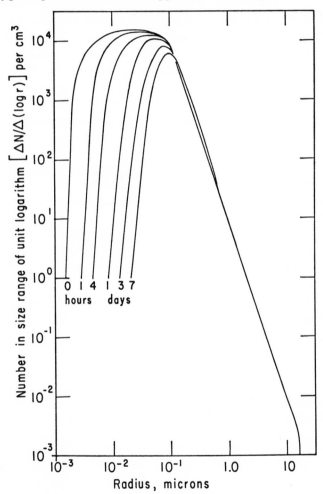

Fig. 4.1.—Size distribution of aerosols averaged from measurements in the vicinity of Frankfurt, Germany. Extensions of curve in lower size ranges show the changed distribution that would result from coagulation in the stated periods of time, assuming no new supply after time zero. (After Junge, 1955, 1958.)

Finally, giant nuclei with radii of several microns are found in the atmosphere. A nucleus of NaCl of 10^{-9} g dry mass (approximately 5μ equivalent spherical radius) becomes a water drop of about 25μ radius at a relative humidity of 99 per cent. This is the size of a small drizzle drop with a terminal fall velocity of about 8 cm sec^{-1}. A salt particle with a dry mass of 10^{-8} g or with 10μ radius would produce a drop of 50μ radius, which would be a large drizzle drop with terminal velocity of about 30 cm sec^{-1}.

Dusts, smokes, and dry haze producing the principal optical scattering in the atmosphere have radii of 0.1 to 1.0μ. Sizes up to 100μ (0.1 mm) or more occur in dust storms and in areas of industrial pollution, but these larger particles settle out fairly rapidly. The common pollens also are in this size range, although some of the allergenic pollens are smaller and travel considerable distances.

In natural fogs and clouds the droplet concentrations are of the order of 100 or more per cm^3. If each droplet corresponds to one nucleus, Figure 4.1 suggests that nuclei in the range of 0.5 to 1.0μ radius are the most important. All evidence points to supersaturations of less than 0.1 per cent in natural liquid-droplet clouds.

Coagulation of Aerosols

The general theory of coagulation of sols was developed by Smoluchowski (1918, 1926). For smokes and general atmospheric aerosols the standard treatment is that of Whytlaw-Gray and Patterson (1932).

The theory is founded on a statement of the rate of collision of particles as they are diffused through the air by Brownian or turbulent motions. The specific flux of matter, that is, the amount moving across a unit area in unit time, is well known in meteorology. It is given by the gradient of concentration of the material along the normal to the area multiplied by a coefficient of diffusion, D. In terms of particle concentrations, N (cm^{-3}) the expression is

$$\frac{1}{A}\frac{dq}{dt} = -D\frac{dN}{dx}, \qquad (4.1)$$

where q is the total number of particles passing through the plane of area A, and x is the perpendicular distance from the plane; D is in the usual units of cm^2 sec^{-1}.

Consider a single isolated sphere with radius S in an environment having an initial concentration of particles N_0. The total flux of particles to the sphere would be

$$\frac{dq}{dt} = 4\pi S^2 D\frac{dN}{dr}, \qquad (4.2)$$

where dr refers to the radial direction outside the sphere. Rearranging terms, generalizing the radius, and assuming a steady rate dq/dt, we may write the integrals

$$\frac{dq}{dt} \int_S^\infty \frac{dr}{r^2} = 4\pi D \int_{N_S}^{N_\infty} dN \,. \tag{4.3}$$

The distribution is such that $N_S = 0$ and $N_\infty = N_0$—the initial distribution—so

$$\frac{dq}{dt} = 4\pi SDN_0 \,. \tag{4.4}$$

If instead of a single sphere we consider that every one of the particles can collide and coagulate with another one, we would have $N/2$ times as many collisions—one collision for each pair—and each collision would decrease N. Equation (4.4) can then be generalized to

$$-\frac{dN}{dt} = 2\pi SDN^2 \,. \tag{4.5}$$

But two colliding particles may not be of the same size, so S becomes $(S_1 + S_2)/2$, and each would diffuse toward the other at its own rate, so D becomes $D_1 + D_2$. The S_1 and S_2 are radii of spheres of influence or capture around the particles. We may now write

$$-\frac{dN}{dt} = \pi (D_1 + D_2)(S_1 + S_2) N^2 \,. \tag{4.6}$$

It has been shown by Einstein (1905) that if we suppose a number of particles to be distributed at random in a fluid, the diffusion coefficient of the particles is

$$D = kTB \,, \tag{4.7}$$

where k is Boltzmann's constant. For particles of radius 1 to 40μ, $B = 1/6\pi\eta r$, the $6\pi\eta r$ being the viscous force at unit speed for a sphere in this size range.[1] For particles comparable in diameter to the mean free path, l of air molecules, a correction factor introduced by Cunningham (1910) is inserted to make

$$B = \frac{1 + (al/r)}{6\pi\eta r} \,. \tag{4.8}$$

At ordinary temperatures l is about 10^{-5} cm while a has been determined by Millikan (1923) to be about 0.9 cm. Thus the correction becomes important for particles with radii in the submicron range, which are the sizes where coagulation becomes significant.

Inserting the value for B from (4.8) into Einstein's equation (4.7), we have

$$D_1 + D_2 = \frac{kT}{6\pi\eta} \left[\frac{1 + (al/r_1)}{r_1} + \frac{1 + (al/r_2)}{r_2} \right] \,. \tag{4.9}$$

[1] See the discussion of the Stokes law in the next section.

Before substituting this expression into (4.6), we may follow the procedure of Whytlaw-Gray and Patterson and write $S_1 = sr_1$, $S_2 = sr_2$. This device merely implies that the sphere of influence is directly proportional to the radius with a proportionality factor s. Then (4.6) becomes

$$-\frac{dN}{dt} = \frac{kTs}{6\eta}\left[\frac{1+(al/r_1)}{r_1} + \frac{1+(al/r_2)}{r_2}\right](r_1+r_2)N^2. \quad (4.10)$$

If the aerosol population is homogeneous, that is, if $r_1 = r_2 = r$, the equation reduces to

$$-\frac{dN}{dt} = \frac{2kTs}{3\eta}\left(1+\frac{al}{r}\right)N^2. \quad (4.11)$$

An example computed from values in this simple equation will serve to illustrate the rates of coagulation. The value $s = 2$ means that the particles coagulate at any contact up to grazing incidence. In this case, with a temperature of $293°$, the pre-parenthetical factor becomes 2.97×10^{-10} cm^3 sec^{-1}. If $r = 0.9 \times 10^{-6}$ cm, $al/r = 10$ and the factor in parentheses is 11. If N is taken as 10^4 cm^{-3}, the coagulation rate becomes 0.327 cm^{-3} sec^{-1} or 19.62 cm^{-3} min^{-1}. If r were 0.9×10^{-5}, the parenthetical factor would be 2 instead of 11, so the coagulation rate would be 18 per cent of that just computed. It is apparent that the smaller the particles and the greater their number density the greater will be the rate of coagulation. From these relations the lower limit of the aerosol size spectrum can be understood.

The nature of the relation (4.10) is such that aerosols must coagulate faster in a population of non-uniform sizes than when they are all of nearly the same size. For example, a particle of radius 10^{-6} cm will coagulate three times as fast with particles of 10^{-5} cm radius as with those of 2×10^{-6} cm radius and 30 times as fast with those of 10^{-4} cm radius. From these examples it is apparent that small aerosols ($r < 10^{-5}$ cm) coagulate rapidly with cloud droplets ($r > 10^{-4}$ cm).

Junge (1955, 1957, 1958) has examined the effects of coagulation in modifying the size spectrum of aerosols in natural situations. His somewhat idealized distribution of aerosols sampled mainly in the vicinity of Frankfurt, Germany, is shown in Figure 4.1. The 3 to 1 logarithmic increase of N with decrease in r is shown. The several curves at the small end of the spectrum show how the initial distribution at 0^h would be modified if coagulation occurred without the addition of new particles to the atmosphere. In the steady state the coagulation would be balanced by production of new particles to preserve the initial distribution.

Two modifications of equation (4.10) are necessary in certain special cases. One case, treated by Greenfield (1957), deals with the coagulation of particulate aerosols with cloud droplets involving N_p particles and N_c droplets. The product N_pN_c replaces the N^2 in (4.5), (4.6), and (4.10). In this case, only the particles are reduced in number by the process, the droplets, with radii of the order of 10μ, being too large to be affected. Equation (4.10) may be written symbolically as

$$-\frac{dN_p}{dt} = K\ N_p N_c ,\qquad (4.12)$$

where K is a "coagulation coefficient" combining the various right-hand terms.

Another modification, following a suggestion of Whytlaw-Gray and Patterson (1932), is to substitute for K a K' which is the sum of $K + K_*$, where K is recognized as characteristic of Brownian motion and K_* of turbulent motion. An expression derived by Smoluchowski gives

$$K_* = \tfrac{4}{3}\frac{\partial u}{\partial n}(r_1 + r_2)^3 ,\qquad (4.13)$$

where $\partial u/\partial n$ is the gradient of air velocity across the streamlines. A velocity gradient of 4 sec^{-1} is suggested by Whytlaw-Gray and Patterson, but a value of 10 sec^{-1} is considered by Tunitzki (1946).

Sedimentation and Fallout

The fall of particles in the atmosphere depends, among other things, on the air density; a particle that may have an appreciable fall speed at 80 km may essentially float at 20 km. Sedimentation, therefore, sometimes results in concentrations in selected layers, especially in the stratosphere where the particles are out of reach of the clouds which produce washout. Fallout implies that the particles reach the surface of the earth under gravity.

Small particles, as well as cloud droplets, have a very small downward speed relative to the air, usually referred to as the terminal velocity. It is the speed attained when the downward inertial force is balanced by the resistance force of the air. The inertial force under gravity in air is the buoyancy force

$$g(M_p - M_a) = \tfrac{4}{3}\pi r^3(\rho_p - \rho_a)g \qquad (4.14)$$

for a sphere of radius r. The masses and densities are M_p, ρ_p and M_a, ρ_a for the particle and the air, respectively.

The resistance force is turbulent in character and requires consideration of the non-dimensional *Reynolds number*

$$Re = 2\rho_a ru/\eta ,\qquad (4.15)$$

where u is the speed of the sphere and η is the dynamic viscosity.[2] Since Re is itself a function of u and r, it is usually the practice to combine it with other parameters to form a function containing only properties of the fluid (air).

---

[2] The subject of Reynolds number theory is omitted in this book. See textbooks in fluid mechanics.

In fluid mechanics texts one finds the derivation of the resistance force on spheres, given in its simplest form as

$$6\pi\eta r u[C_D Re/24] , \qquad (4.16)$$

where C_D is the drag coefficient, also non-dimensional. When we equate this force to the inertial force (4.14), we have

$$6\pi\eta r u[C_D Re/24] = \tfrac{4}{3}\pi r^3(\rho_p - \rho_a)g , \qquad (4.17)$$

and, denoting u as u_T, the terminal velocity, we find

$$u_T = \tfrac{2}{9}\frac{\rho_p - \rho_a}{\eta} g r^2 \frac{(24)}{(C_D Re)} . \qquad (4.18)$$

For particles small enough ($r < 40\mu$ for unit density) to have appreciable residence times in the atmosphere, $C_D Re = 24$ and the part in parenthesis is

TABLE 4.1

REYNOLDS NUMBER, Re, AND DRAG COEFFICIENT, C_D, FOR SPHERES OF VARIOUS RADII AND UNIT DEN- SITY FALLING IN AIR AT 20° C AND 760 MM HG

Radius, cm	Re	C_D
10^{-5}.........	1.7×10^{-8}	1.412×10^9
10^{-4}.........	1.7×10^{-5}	1.412×10^6
5×10^{-4}.........	2.125×10^{-3}	1.129×10^4
10^{-3}.........	0.017	1.412×10^3
5×10^{-3}.........	1.80	15.0
10^{-2}.........	9.61	4.2
5×10^{-2}.........	269	0.671
10^{-1}.........	866	0.517
2×10^{-1}.........	$2,357$	0.559

unity. The expression then reduces to the form developed by Stokes (1850), and the small particles are said to be in the Stokes law range of sizes. For larger spheres, $C_D Re$ must be supplied as an appropriate number, best determined experimentally.

When the expression for Re given in equation (4.15) is substituted in equation (4.17), we find

$$u_T^2 = \tfrac{8}{3}\frac{g}{C_D}\frac{\rho_p - \rho_a}{\rho_a} r , \qquad (4.19)$$

which is true for situations in which $Re > 1$, but this expression does not help much since C_D has to be determined in each case. It shows that outside the Stokes region the terminal velocity is no longer a function of r^2.

Some values of C_D and Re are given in Table 4.1 for particles of unit density having various radii. Davies (1945) determined that for a Reynolds number of 0.82 the Stokes law gave a value of terminal velocity too high by 10%; for

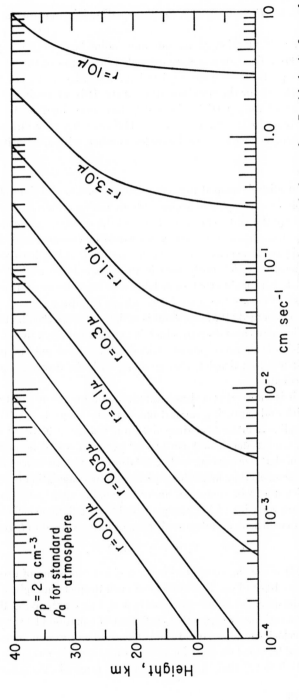

FIG. 4.2.—Terminal velocity of spherical particles of various radii as a function of height in a standard atmosphere. Particle density 2 g cm⁻³. (After Junge, Chagnon, and Manson, 1961.)

$Re = 0.38$, 5%; for $Re = 0.15$, 2%; for $Re = 0.074$, 1%; and for $Re = 0.037$, 0.5% too high.

Values of u_T as a function of altitude and radius have been computed by Junge et al. (1961). Their results were presented in a series of curves shown in Figure 4.2. The particles were assumed to have a density of 2 g cm^{-3} and the air density was taken for the standard atmosphere. It is interesting to note, for example, that at 40 km a particle of $r = 0.3\mu$ has approximately the same fall velocity as for one of $r = 3.0\mu$ at sea level. The variation with altitude is due to the variation of viscosity, η and Reynolds number, with air density.

Washout

Theoretical studies developed indirectly from experimental data show that the true aerosols are not captured appreciably by falling raindrops. They are so small that they are thought to be carried around the raindrop in the air stream. Giant particles that are already settling out would be captured and thus have their earthward passage accelerated, and large materials of low density—spores, pollens, and the like—which fairly float in the air would be washed out (McDonald, 1962). It is probable that snowflakes scavenge efficiently from the atmosphere aerosols that are too small for raindrops to capture.

As shown by Greenfield (1957) washout is mainly accomplished by the coagulation of particles with *cloud droplets* which in turn are collected by falling raindrops in a collision-coalescence process. Cloud-condensation nuclei, apparently present at least one to a droplet, also go along with the droplets which they originally nucleated.

In chapter 6 it is shown that a sizable raindrop obtains nearly all of its water by colliding with and capturing small droplets in the cloud. It takes a million droplets of 5μ radius to make a raindrop of 0.5 mm radius: $(5 \times 10^{-2})^3 = 10^6 \times (5 \times 10^{-4})^3$. A rainfall depth of 1 cm (1 cm^3 per cm^2) would be made up of 2000 raindrops of that size per cm^2, or 2×10^9 droplets with their accompanying nuclei plus any aerosols they might have picked up by coagulation in the cloud.

The cloud pickup can be computed under reasonable assumptions by means of the coagulation equation (4.12), using a suitable coagulation coefficient. The equation with a coefficient K' integrated from an initial time 0 to t, gives

$$N_p(t) = N_p(0) \exp\left(-K'N_c t\right). \tag{4.20}$$

Greenfield (1957) made the computation for a cloud consisting of droplets all of the same size—10μ radius. The number of such droplets in a given volume followed from the assumed liquid-water content of 4 μg cm^{-3}, giving 100 droplets per cm^3. The turbulent coagulation coefficient of equation (4.13) was based on an extreme velocity gradient of 30 sec^{-1}. The result, expressed as fraction of particles collected by cloud droplets for various durations of coexistence is given in Figure 4.3. It is noted that in the lower Aitken range the coagulation pro-

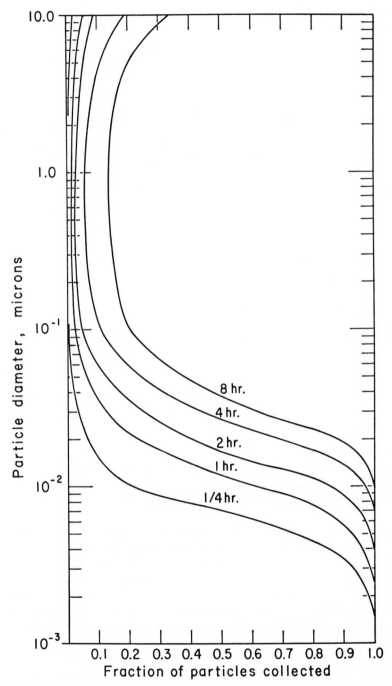

FIG. 4.3.—Collection of particles by cloud droplets coexisting for various periods of time. (After Greenfield, 1957.)

ceeds rapidly. The minimum rate is seen in the cloud-nuclei sizes. However, if every droplet grows around one such nucleus and there are 100 droplets per cm³, most of the nuclei of this size at normal number concentrations are already in the droplets.

The droplets, with the particles they have condensed upon and those they have picked up by coagulation, are swept earthward by raindrops which collect the droplets on the way down. The efficiency of raindrops in collecting cloud droplets is discussed in chapter 6. The efficiency is essentially a function of the sizes of the collecting and collected spheres, but for many common distributions the raindrops collect nearly all of the cloud droplets in their path.

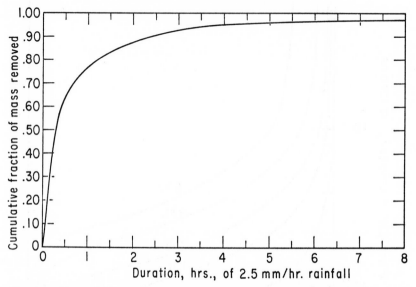

Fig. 4.4.—Removal of mass of aerosol by rain washout in 2.5 mm-per-hour rain occurring after one hour of particle residence in cloud. (After Greenfield, 1957.)

For reasonable values of the various parameters, using a distribution of raindrop sizes as a function of rainfall rate determined by Best (1950), Greenfield found that if the coagulation operated between particles and 20μ droplets as shown in Figure 4.4 for one hour, then if 2.5 mm (0.01 in.) of rain fell in the next hour, the mass of aerosol material washed out of a given volume would be between 75 and 80 per cent of the initial mass. Rains of greater intensity or duration would produce greater washout. Since big rains come from deep clouds, the volume of cloud scavenged would be greater in such cases.

Sources of Nuclei

Particles capable of serving as condensation nuclei come from several sources: (1) fine dusts eroded from the land by the wind; (2) natural combustion (forest, brush, and grass fires) and man-made combustion; (3) detachment from the sea surface by spray or bubbles; (4) volcanic eruptions and man-made explosions; (5) extraterrestrial; (6) nucleogenic-chemical and photochemical reactions. Another process whose role is not completely understood (O'Connor *et al.*, 1959) is the production of nuclei at heated surfaces. Twomey (1960) found that large numbers of nuclei are produced from a soil surface when heated under dry conditions.

The chemical composition of the particles reflects their source and the chemical processes to which they were subjected in the atmosphere. Sulfates, chlorides, and nitrates are common, either in the acid form or with sodium, potassium, calcium, a surprisingly high frequency of occurrence with ammonium ion, and a great variety of other substances. Sulfates are formed by combustion and other chemical processes, as also are nitrates. Chlorides come both from the sea and the land. Ammonia is familiar in decaying organic matter.

A number of complex oxidation processes occur. Haagen-Smit (1956) finds in the polluted air of Los Angeles that ozone is formed in a photochemical oxidation process of hydrocarbons in the presence of oxides of nitrogen. Through the ozone the nitric oxide and sulfur dioxide from combustion are converted to higher oxidation stages and through subsequent reactions with alkaline material are found as nitrates and sulfates. Coste and Wright (1935) suggested that the particles of sulfuric acid commonly found in the atmosphere are formed in a reaction between sulfur dioxide and nitrates. Junge (1958) accounts for the high frequency of occurrence of ammonium sulfate particles by assuming a reaction between ammonia and sulfuric acid. Another problem has been to account for the preponderance of sulfates even over ocean areas where combustion is not important. At shore lines and elsewhere over the oceans there is a considerable amount of H_2S, and Junge (1960) supposes that through stages of oxidation, perhaps with ozone, a conversion to H_2SO_4 results.

Fuels such as coal and oil contain appreciable amounts of sulfur which, upon combustion, produce SO_2. As was first recognized by Aitken (1912), the SO_2 is ozidized to SO_3 in the atmosphere. In the presence of water the SO_3 immediately forms sulfuric acid particles which are highly hygroscopic. Gerhard and Johnstone (1955) studied in detail the photochemical oxidation of SO_2 in air and found that in the presence of sunlight the SO_2 is oxidized to SO_3 at a relatively rapid rate.

Much of the natural sulfate in the atmosphere is produced through reaction with naturally produced H_2S. Conway (1943) reported that H_2S is produced in large quantities by algae in the littoral zone of the sea.

Verzár and Evans (1959) studied the effect of sunlight on nuclei by putting trace gases in filtered, nuclei-free air pumped into polyethylene balloons. They found that H_2S in low concentrations produced tremendous numbers of nuclei

79

in the balloon when exposed to sunlight. Traces of NH_3 also were active in the sunlight but not as much so as the H_2S.

Interesting demonstrations of the nucleating ability of aromatic vapors given off by growing plants have been given by Went (1960). From pine forests, sagebrush, and the like are emitted volatile organic substances, generally classed as terpines, which, especially when acted upon by sunlight, appear as active condensation nuclei.

From all available data, Junge (1960) estimated the residence times of some aerosols in the atmosphere. He found the natural SO_2 derived from the oxidation of H_2S to have an atmospheric residence time of about 15 to 30 days as against a 5-day residence time for industrial SO_2. On this basis Fenn (1960) concluded that the aerosol which he measured on the Greenland icecap, which was predominately sulfate, was of natural origin rather than industrial.

Sea-salt nuclei are left suspended in the air by the evaporation of small droplets detached from the sea in spray, foam, and bursting bubbles. Formerly it was thought that spray was the important source, but now it is known that whitecaps and even inconspicuous, tiny wavelets entrain air to form bubbles near the water surface and that these bubbles are continually bursting to produce droplets in the air immediately above the surface. If the droplets have diameters less than 20 or 30μ they will remain airborne long enough to be evaporated to aerosols.

Several workers have carried out wind-flume experiments to produce bubbles and study them in detail. Notable among these experiments have been those of Woodcock (1953), Woodcock *et al.* (1953), Knelman, Dombrowski, and Newitt (1954), Mason (1955), Blanchard and Woodcock (1957), Hayama and Toba (1958), Isono (1959), and Toba (1959, 1961). Photographs, some taken at very high speeds, of these experiments show that there are two different phenomena which produce droplets from a bursting bubble. One is the development of a jet of water which breaks into droplets before it settles back to the surface. The other is the shattering of the film of the bubble. Experiments indicate that the latter process is the most important for producing droplets of a size that can be sustained in the air.

As might be expected, the rate of droplet production is highly dependent on the wind speed. Toba (1961) determined the number of droplets of less than 50μ diameter formed in a wind flume and found that at a wind speed of 12.1 m sec^{-1} there were more than 40 times as many as were formed at a wind speed of 8.7 m sec^{-1}. Moore and Mason (1954) found that, with a wind speed equivalent to 16 m sec^{-1} at a height of 10 m above the surface, nuclei of mass greater than 2×10^{-13} g were formed at a rate of 40 cm^{-2} sec^{-1}.

To study the production of much smaller nuclei, Mason caused bubbles to burst in cloud chambers. Clean, essentially aerosol-free air was passed into two chambers, one containing sea water and the other distilled water. After an equal number of bubbles had burst in each chamber, an expansion was made; a dense cloud was then observed above the sea water but not above the distilled water. It was determined that bubbles of 3 mm diameter produced between 100 and 200 nuclei, the majority of which were estimated to have salt contents between 10^{-15} and 2×10^{-14} g.

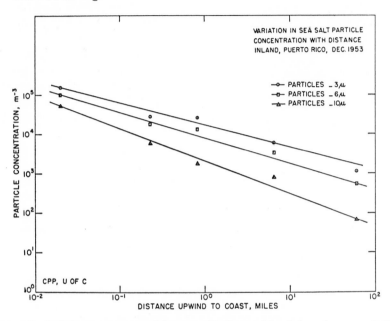

FIG. 4.5.—Lodge's measurements of decrease with distance inland of number concentration of giant chloride particles in Puerto Rico.

Land areas are greater sources of Aitken nuclei than the oceans. O'Connor (1961) obtained data from a high headland on the western coast of Ireland showing the following relation to wind direction:

Direction.....	N	NE	E	SE	S	SW	W	NW
N/cm³........	2610	5020	3850	1280	1160	690	890	5050

Maritime air comes from SE, S, SW, and W at that location; the other directions bring land air. Counts as low as 100 or less per cm³ have been reported on numerous occasions over calm seas in mid-ocean.

Sea-salt nuclei in the large and giant sizes have been identified and counted at various distances from the shore line in Puerto Rico by Lodge (1955). The particles were chemically identified by their chloride ion. The results, shown in Figure 4.5, suggest that many large sea-salt nuclei are thrown into the air by the surf and that these decrease rapidly in number with distance inland. Gravitation slowly removes the larger ones, but evidence from other investigations indicates that trees, bushes, and other vegetation capture many of the particles in the low levels. At inland locations Byers *et al.* (1957) found that there were more of these large and giant chloride particles aloft than at the ground.

Of the easily recognized chemical components, Junge (1958) considers that the continents contribute most of the NH_4^+, NO_3^-, NO_2^-, SO_4^{--}, and Ca^{++}, while the oceans are the main source of Cl^-, Na^+, K^+, and Mg^{++}.

A useful way of determining the broad distributions, and thereby deducing the sources of these components, is by the analysis of rain water. One may assume that the rain and snow wash out a representative aerosol sample of at least the lower half of the troposphere. The chemical composition of rain water is of interest to agricultural chemists and workers in other sciences, so it has

TABLE 4.2

TYPICAL VALUES OF CHEMICAL COMPONENTS IN
RAINS IN THE UNITED STATES
(MG/LITER)

Ions	Continental Interior	Sea Coast
Na^+.	0.3	5.0
Ca^{++}.	3.0	0.3
NH_4^+.	0.2	0.02
Cl^-.	0.2	9.0
SO_4^{--}	3.0	1.0*

* Non-industrial coastal areas.

been possible to organize, at least in Europe and North America, a network of stations. The data have been published regularly in the Swedish journal, *Tellus*. From these observations it is possible to draw maps of various continents or regions showing isopleths of the components by months or seasons. Some typical values for rains in the United States taken from maps prepared by Junge and Werby (1958) are given in Table 4.2.

In two papers Eriksson (1959, 1960) has summarized the global ocean, land, and atmospheric cycle of chlorine and sulfur compounds. The sulfate, etc., has been expressed in sulfur equivalent. Lodge (1960) interpreted the figures in terms of the ratio of chlorine to sulfur in the atmospheric load. From Eriksson's two papers the figures show, respectively, (Cl) = 1.43(S) and (Cl) = 1.75(S). Lodge had at hand only the first ratio and took it to represent the mean atmosphere. Sea water was taken as the upper limit of the (Cl)/(S) ratio. The sulfate content of sea water is $\frac{1}{7}$ that of chloride, and since the sulfur is by weight $\frac{1}{3}$ of the sulfate mole, (Cl) = 21(S) in sea water.

On logarithmic paper on which the lines for the relationship in the sea and in the mean atmosphere were entered, Lodge plotted observations for various stations. For each location a line was fitted to the observations by the least-squares method, producing in each case a regression of the form (Cl) = $a +$ (S)^b. Several such locations are represented in Figure 4.6. Weather-ship station "November," which is located in the calm part of the Pacific about halfway be-

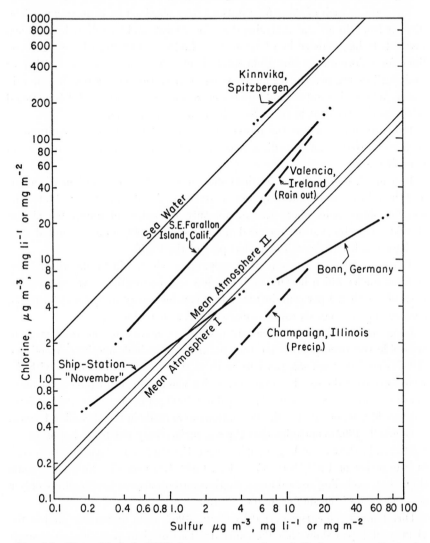

Fig. 4.6.—Plots of Lodge's (1960) data on concentrations of chlorine versus sulfur in air samples (μg m^{-3}), precipitation analysis (mg liter^{-1}), and in rainout (mg m^{-2}) compared with sea water and with Eriksson's mean atmospheres. Where source of measurement is not indicated, data are for air samples.

83

tween San Francisco and Honolulu, exhibits a ratio approximating that of the mean atmosphere. Kinnvika, Spitzbergen, representing the stormy part of the Arctic Ocean, shows essentially a sea-water regression line, while at Bonn, Germany, the (S) component dominates. At southeast Farallon Island, lying thirty miles outside the Golden Gate, a ratio about midway between that of the mean atmosphere and that for sea water is found.

The data at each of these four stations were obtained by capturing particles directly from the air and analyzing them for chloride and for sulfate by chemical methods to be discussed in a later section. Lodge also computed the regression lines for chlorine *vs.* sulfur from analysis of rain water, expressed either as "rainout" in mg per square meter or as precipitation in mg per liter of rain water. Where air samples and rain water samples were both used, Lodge found fairly good agreement in (Cl)/(S) ratio between the methods.

It is difficult to assess the representativeness of true marine conditions at stations located on an immediate coast where the surf produces a local excess of sea-salt nuclei.

In general, the maritime locations show less total aerosol content of the air and the rain water than do continental stations. The data in Figure 4.6 for the weather ship "November" exhibit not only low values of sulfate but also of chloride. The observations included only the summer, when uninterrupted anticyclonic conditions prevail over that part of the Pacific.

Twomey (1959), by means of an apparatus which will be described later, counted nuclei which became active at low supersaturations. At supersaturations of less than 1 per cent, he found that continental air at Sydney, Australia, contained more than six times as many nuclei as maritime air.

Large and giant nuclei obviously contribute heavily to the total mass of aerosol in the atmosphere when they occur in appreciable numbers. Woodcock (1953) found that sea-salt particles in these sizes increase in numbers with the wind force over the sea. For example, with a wind of Beaufort force 3 the number of sea-salt particles of 10^{-10} g was about 7000 per cubic meter while at force 7 it was 50,000 per m³. The determinations were made in the Hawaiian Islands.

Gambell (1962) concludes that the sea surface may not be as important as land even in the case of hygroscopic nuclei. His studies are based on the ratios in rain water of Ca^{++}/Na^{+}, Ca^{++}/Cl^{-}, Ca^{++}/K^{+}, and Cl^{-}/Na^{+}. His results show that, excluding coastal areas, marine aerosols appear to contribute only a small portion of the soluble aerosols in rain.

Direct measurements reported by Okita (1962) from flights over and in the vicinity of Japan show that the concentrations of giant particles are greatest over the sea and along the coast and decrease to $\frac{1}{10}$ at distances of about 100 km from the coast.

Vertical Distribution

Since the preponderant mass of aerosol material originates at the surface of the earth, it is natural that in any state of diffusion, transport, fallout, and washout the amount of aerosol would decrease with height.

For any lapse rate less than autoconvective (less than 3.41° C per 100 m) the volume occupied by a mole or gram of air increases with height; therefore, the number density of particles would decrease with height due to that effect alone. To avoid the air-density effect, it is desirable to define the particle concentration in terms of a mixing ratio expressible either as a mass ratio or as the ratio of the number of particles to the number of molecules of air in the same volume, and designated as ν. The fractional change of this quantity per unit-of-height increase would have a variable negative value, which is conveniently given as a reciprocal, $-1/\lambda$, such that

$$\frac{1}{\nu}\frac{d\nu}{dz} = -\frac{1}{\lambda}. \tag{4.21}$$

Integrated from the ground at $z = 0$ to a height z the distribution becomes

$$\nu = \nu_0 \exp\left(-z/\lambda\right), \tag{4.22}$$

wherein it is seen that λ has the dimensions of a length and is, in fact, the height at which ν has the value $1/e$th of that at the ground. But λ may also be expressed as

$$\lambda = D/(u_T - U), \tag{4.23}$$

where u_T is the terminal velocity of the particles, U is the updraft speed of the air, and D is the diffusivity, here eddy diffusivity, for particles. Then

$$\nu = \nu_0 \exp\left[(U - u_T)z/D\right]. \tag{4.24}$$

In still air, since u_T increases with size of the particles, the value of the exponent must decrease with particle size. In updrafts the particles would be carried upward, and in downdrafts the exponential decrease with height would be accentuated. (Note that U is positive upward and u_T is positive downward.) The turbulent diffusion or mixing is strongly dependent on the thermal stability, being sharply reduced in temperature inversions and other stable layers. Subsidence of the air sharpens the layering of the distribution, while convective penetrations cause sudden changes at critical heights.

Equation (4.24) does not take into account sources and sinks in the atmosphere. Clouds temporarily or permanently modify the distribution. If the cloud droplets have collided and coalesced with each other, their combined nucleating material may appear as larger but fewer nuclei after the droplets have evaporated (Eriksson, 1959). If there is a growth to raindrops, washout, as described on previous pages, will provide a sink at the cloud levels. Sources of particles other

85

than ground sources include interplanetary sources, such as meteoric dust which filters through the atmosphere, and particle production by certain gas reactions, often photochemical, such as in the ozone layer.

From airplane flights one notices the pronounced layering of haze, marking inversions, stable layers, or limits of convective penetration. In a series of flights in which they measured the concentration of Aitken particles, Sagalyn and Faucher (1956) found a predominant exchange layer in the lowest 5000 feet or so with a relatively slow exponential decrease, above which the concentration decreased sharply to quite low values. Subsidence rates could be calculated from measurements taken over several hours (Byers, 1957).

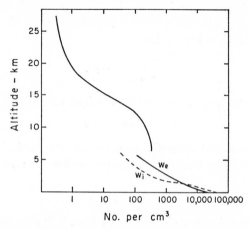

Fig. 4.7.—Vertical distribution of Aitken particles. Lower: Wi, Wigand (1919), manned balloon flights; We, Weickmann (1957), airplane flights, smoothed curve. Upper: Junge (1961), free balloon flights, smoothed curve.

Aitken particles.—Average vertical distributions of Aitken particles ($<0.1\mu$ radius) are shown in Figure 4.7. The measurements of Wigand (1919), labeled Wi in the figure, were averaged from fourteen flights by manned balloon over Europe. The data of Weickmann (1957), labeled We, the mean of twelve airplane flights, have been entered in the form of a smoothed curve. Wigand's lower values aloft probably are biased by the selection of anticyclonic conditions for the manned balloons, a precaution that was not necessary in Weickmann's airplane flights. The curve for heights above 5 km is smoothed from the data reported by Junge (1961) from seven flights of recoverable free balloons carrying automatic Aitken counters. The curve shows very little decrease with height in the upper troposphere, then an approximately exponential decrease to around 17 or 18 km, with a much slower decrease in the 20–30 km range. Photochemical source layers in the upper troposphere and in the stratosphere are suggested by this curve. These source layers are more apparent for larger nuclei, discussed on a succeeding page.

Kumai and Francis (1962) examined in the electron microscope the aerosol residues left by 356 snow crystals after they had sublimed to vapor. The snow crystals were collected from summer snow occurrences on the Greenland ice-cap. The snow crystals apparently bring down to the ground a representative sample of the aerosols from somewhere in the clouds. The greatest concentration was in the size range 0.005 to 0.02μ diameter, well within the Aitken sizes. The threshold of detection was 0.003μ. A relatively large substance a few tenths to a few microns in diameter was found in the center of nearly all of the snow crystals —apparently the ice-nucleating particle. These latter are discussed in a subsequent section.

Large nuclei.—Large nuclei ($0.1 < r < 1.0\mu$) and giant nuclei ($r > 1\mu$) are counted by more laborious techniques than the Aitken particles, and hence their concentrations have not been as thoroughly surveyed. Balanced against this disadvantage is the fact that after capture they can be studied individually and their chemical compositions relatively easily determined.

Concentrations in these size ranges in the free air are one to three orders of magnitude less than in the Aitken range, yet these particles represent a greater mass load than the Aitken particles. They are the predominant natural condensation nuclei, many being in the form of hygroscopic, soluble salts.

Fenn (1960) studied the large nuclei on the Greenland icecap and found that sulfates predominate. The size spectrum showed two very narrow maxima of 1.6 cm^{-3} at 0.5μ radius and 1.4 cm^{-3} at 0.7μ radius. The concentration is less by a factor of 100 in the night of winter, which supports the idea of photochemical oxidation of SO_2 to SO_3 and subsequent conversion to H_2SO_4.

During the 1958 to 1960 moratorium on tests of atomic weapons the U.S. Department of Defense and Atomic Energy Commission carried out an urgent program of surveying the "natural" aerosol content of the stratosphere. As a result there now exists more knowledge of the aerosol load above 10 km than in the main part of the troposphere. The high-altitude balloon flights which measured the Aitken particles as shown in Figure 4.7 also included measurements in the sizes of $0.1 < r < 1.0\mu$ by means of special impactors and other collectors. U-2 airplanes, flying mostly at 20 km, covered a large range of latitude from the topics to high latitudes in both hemispheres. The analysis of the data was placed in the hands of Junge *et al.* (1961).

The results showed that although the number concentration of Aitken particles decreases with height through the stratosphere, the large particles, $0.1 < r < 1.0\mu$, have a pronounced maximum in concentration at an altitude of about 20 km. This distribution is seen in Figure 4.8, representing a "typical" profile presented by Junge and Manson (1961) and Chagnon and Junge (1961).

The chemical composition of the particles was determined by electron microbeam probe analysis and by X-ray fluorescence vacuum spectrometry. The results showed that 89 per cent by weight of the material identified in this way was sulfate.[3] A chemical reagent test for NH_4^+ showed a concentration in agreement with that calculated stoichiometrically from the SO_4^{--}. There was evidence that some of the sulfate was from sulfuric acid. Evidently the stratospheric layer of large aerosols is primarily ammonium sulfate, $(NH_4)_2SO_4$, with some H_2SO_4.

In explaining the presence of the stratospheric maximum, one naturally interprets the shape of the profile, with its relatively lower values at the tropopause, to mean that the particles do not penetrate directly from the tropo-

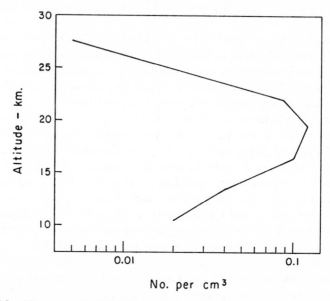

FIG. 4.8.—"Representative" distribution given by Junge and Manson (1961) of large particles—$0.1\mu < r < 1.0\mu$, approximately—above 10 km.

sphere into the stratosphere either by vertical or horizontal mixing through tropopause breaks. The chemical composition of the particles apparently precludes the possibility of extraterrestrial origin. Junge and Manson (1961) suggested that the aerosols are formed in the ozone layer, with which the maximum roughly coincides, by oxidation of trace gases diffused upward from the troposphere. The compounds H_2S and SO_2 are considered to be the most likely particulate-forming gases in this process. Growth by "condensation" of the oxidized gas is assumed. Junge and Manson argue that since troposphere con-

[3] Actually the sulfur itself was detected by the methods used, but stoichiometric considerations indicated that it was present as sulfate.

centrations of gaseous sulfur components of the order of 10^{-6} g m^{-3} are found, stratospheric sulfur concentrations of the order of 10^{-9} to 10^{-8} g m^{-3}, which they observed in particles, are not surprising. Sedimentation removes particles of radius appreciably larger than 1μ. At 20 km the fall speed of a sphere of radius 1μ, density 2 g cm^{-3}, is the same as that of a similar sphere of radius 2μ at sea level—about 0.07 cm sec^{-1} or 60.5 m day^{-1}.

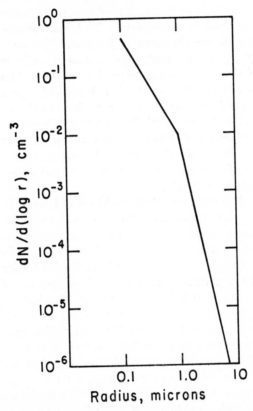

Fig. 4.9.—Average smoothed curve of size distributions of large and giant particles measured by Junge and Manson (1961) at about 20 km.

The size distribution at about 20 km of the collections used in the vertical profiles is shown in Figure 4.9 from Junge and Manson. The two types of collections—balloon and aircraft—produce radically different curves at sizes around $r = 0.1\mu$ and smaller, and at these sizes it is obvious that collection problems make it difficult to arrive at appropriate concentrations. The sharp decrease in concentrations for $r > 1\mu$ is very striking; it is interpreted by the authors to mean that particles of a different origin—extraterrestrial—are essentially the only ones present in that size range. On five aircraft flights at approximately 20 km the average $dN/d(\log r)$ values were less than 10^{-4} cm^{-3} for $r = 3\mu$, about 10^{-5} for 5μ, and of the order of 10^{-6} for 10μ particles.

Newkirk and Eddy (1964) used a balloon-borne instrument for measuring the sky radiance and from that obtained the size distribution of aerosols above the balloon by solving the equation for transfer of radiation through a turbid atmosphere. From two such flights they found a sharp maximum in concentration of particles of $r = 0.15\mu$ at 17 to 18 km, thus adding strong independent evidence of the stratospheric peak in concentration. Their particle-size distribution at 20 km also fitted quite well the curves of Junge and Manson.

Noctilucent clouds, observed during the summer in northern latitudes, have optical characteristics corresponding to particles of 0.1 to 1.0μ radius in concentrations of the order of 10^{-2} to 1 cm^{-3}. It is generally agreed that they are not water or ice particles. The clouds are confined to the relatively narrow altitude range of 75 to 85 km. At this height is found the essentially isothermal layer of absolute minimum of atmospheric temperature, above the mesosphere.

Dust layers are commonly observed in the stratosphere, expecially after volcanic eruptions. Those that cannot be traced to volcanoes or bomb debris could be of interplanetary origin.

Giant nuclei.—Two classes of giant nuclei are of interest in cloud physics, one originating by detachment from the surface of the earth and the other streaming into the atmosphere from space. The former class can nucleate condensation of drops large enough to constitute the beginnings of rain, and the latter class is thought to be important in nucleating the ice phase.

The giant nuclei of terrestrial origin are easily caught on various impactors, sieves, etc., and studied individually as to size, chemical composition, and, with metering of the air from which they are caught, as to number concentration. In most cases each chemical is identified and studied separately.

In Figure 4.10 are shown some vertical distributions of giant chloride particles, with a set of data on sulfate included. The plots, from Lodge (1955) and Byers *et al.* (1957), are from airplane flights over the ocean in the vicinity of Puerto Rico and over the continental United States in the middle and lower Mississippi Valley. Only particles of radius $\gtrless 1.5\mu$, computed as spherical NaCl in the case of the chlorides and sulfate particles of $r \gtrless 2.7\mu$ (spherical H$_2$SO$_4$), are counted.

The Puerto Rico flights show a steady decrease in concentration of giant chlorides up to about the level of the base of the characteristic trade-wind cumulus clouds of the region, then a nearly constant value from there up to the trade-wind inversion.[4] A marked decrease occurs above the trade-wind inversion. Woodcock (1953) noted the same type of distribution in the region of the trade winds near the Hawaiian Islands.

[4] For a meteorological description of the trade-wind belt see Riehl (1954).

Over the continent the giant chloride particles are quite sparse at the ground, and even at 300 m are two orders of magnitude fewer in numbers than in Puerto Rico, but above 300 m there is very little change with height, so that above about 2500 m the counts are about the same as Puerto Rico. Seven level flights at various heights up to about 4600 m in the vicinity of Tucson, Arizona—point marked *TUS* on the graph—gave counts comparable with those of the Middle West.

The sulfate counts were made on a level flight at about 2500 m over the Mississippi Valley between Central Illinois and the Gulf of Mexico. The concentra-

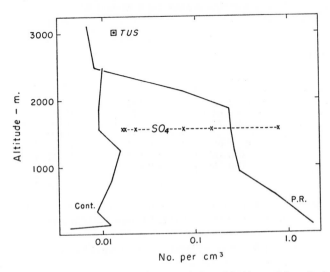

Fig. 4.10.—Average of collections on Millipore of giant chloride particles, $r \geq 1.5\mu$, on five flights near Puerto Rico (*P.R.*), and on twelve horizontal and vertical flights in the Middle West (*Cont.*) during the drought summer of 1954. Sulfate measured on one long flight in the lower Mississippi Valley at about 1500 m shown by horizontal dashed line. Average of giant chloride particles on seven essentially horizontal flights in February, 1956, at various altitudes to 4600 m near Tucson, Arizona, shown by point labeled *TUS* at 3000 m.

tion of giant sulfate particles exceeded the average chloride counts for the continental area by factors of two to a hundred. Some of the highest counts were found over the waters of the Gulf of Mexico.

From these plots one might conclude that from about 2.5 km to perhaps 5 km there is a normal atmospheric giant chloride-particle concentration of about 0.01 cm^{-3} over both continents and oceans. Giant chloride particles in these concentrations could be of both continental and maritime origin. A general background of giant sulfate particles varying from 0.01 to 1 cm^{-3} might be expected in the middle troposphere over most of the earth.

If the sulfates and chlorides are added together and allowance is made for other ions in soluble salts, such as NO_3^-, CO_3^{--}, various oxides, organics, and insoluble particles, one might expect the total count of giant particles to be of the order of 1 to 10 cm^{-3} with locally extreme values in sea surf, volcanic eruptions, smokes, and dust storms.

On the research flights of the British Meteorological Office chloride particles of large and giant size (mass greater than 10^{-13} g NaCl, $r > 0.2\mu$) have been

TABLE 4.3

A. RANGE IN NUMBER CONCENTRATIONS PER CM^3 OF CHLORIDE PARTICLES
IN TEN AIRPLANE ASCENTS OVER ENGLAND IN 1957

(From data of Durbin and White [1961])

ALTITUDE (feet)	MASS AND EQUIVALENT SPHERICAL RADIUS AS NaCl			
	$> 10^{-13}$ g $> 0.18\mu$	$> 10^{-11}$ g $> 1.03\mu$	$> 10^{-10}$ g $> 2.2\mu$	$> 10^{-9}$ g $> 4.8\mu$
100.............	0.042–2.32	0–0.854	0–0.235	0–0.036
200.............	0.024–1.64	0–0.776	0–0.253	0–0.042
500.............	0.024–1.74	0–0.516	0–0.094	0–0.019
1000.............	0.033–1.25	0.003–0.335	0–0.022	0–0.004*
2000.............	0.012–1.06	0.006–0.231	0–0.034	0–0.006*
5000.............	0–0.165	0–0.079	0–0.012	0
10,000............	0–0.343	0–0.002	0	0

B. NUMBER CONCENTRATIONS PER CM^3 OF CHLORIDE PARTICLES IN SEVEN
FLIGHTS AT 1000 TO 3000 FEET OVER BRITISH ISLES
AND ADJACENT SEAS IN 1957

(From data of Singleton and Durbin [1962])

	MASS AND EQUIVALENT SPHERICAL RADIUS AS NaCl			
	10^{-13} g 0.18μ	10^{-11} g 1.03μ	10^{-10} g 2.2μ	10^{-9} g 4.8μ
Means of sampling periods..........	0.030–2.3	0.005–0.78	0.001–0.21	<0.001–0.017
Maxima of 7 flights ..	3.1	2.8	0.82	0.075

* Interpolated.

sampled by methods similar to those of Lodge. The results have been reported in papers by Durbin and White (1961) and Singleton and Durbin (1962). Table 4.3 shows data of ten airplane ascents over England and seven horizontal low-level flights over the British Isles and adjacent seas. In general the counts of the giant chlorides represent values between those of the U.S. continental and Puerto Rican flights. The sampling periods in the horizontal flights varied in length between 15 and 50 minutes. They show higher values than those obtained in the vertical ascents.

Vertical distribution in the upper atmosphere.—Concentrated in the zodiacal zone within the solar system is a cloud of dust particles long known as responsible for the observed zodiacal light. The earth collects this interplanetary dust, much of which occurs in particles of giant size. They are small enough, however, to have fall velocities in the range of millimeters to a few centimeters or so per second and thus have residence times above the tropopause—out of the reach of washout processes—of 10 to 100 days. Other large magnetic spherules are found in ocean-bottom sediments. They are identified as either particles that have melted in the fiery ablation of meteors (Buddhue, 1950) or those particles of the zodiacal dust cloud of such a size that in entering the atmosphere they are heated just enough to melt but not enough to vaporize (Öpik, 1956). Particles of radius 5 to 30μ are of this kind. They are only transitory in the atmosphere. Our interest is in particles of radius 1.0 to 10μ whose masses, assuming a density of 2 g cm^{-3}, would be about 10^{-11} to 10^{-8} g.

Measurements from rockets and satellites now furnish data on the zodiacal dust. The records considered to be most reliable are those obtained from microphone type dust-particle sensors in which a microphone is attached to a metallic sounding board. An impacting particle gives a mechanical impulse to the sounding board which generates an electrical signal. The signal is amplified and the pulse height is proportional to the particle mass for any assumed particle speed relative to the space vehicle, about 30 km sec^{-1} when in orbit.

The number of dust-particle impacts is found to be quite variable over short periods of time, especially during meteor showers, but an average cumulative-mass-distribution curve of particle influx has been presented by Dubin and McCracken (1962), McCracken *et al.* (1961), Alexander *et al.* (1962) based on all available data of U.S. and Soviet satellites and rockets. The curve is shown in Figure 4.11. It indicates that particles in the giant size, 10^{-11} g and larger, produce several impacts per square meter per second.

Carleton (1962) pointed out that the dust particles are actually in orbit around the earth, so a relative velocity of 15 km sec^{-1} is more reasonable than the 30 km sec^{-1} for determining the influx to the earth. He computed the influx rate at the top of the atmosphere to be 1/600 of the rate of influx on satellites. The total accretion to the earth is figured at 260 metric tons per day.

Observations of twilight scattering were used by Volz and Goody (1962) to deduce the distribution of dust particles in the upper atmosphere. A maximum of turbidity was found between 15 and 30 km, consistent with the findings of Junge *et al.* Dust was nearly always present at all levels up to 65 km, and above 30 km there was a nearly constant mixing ratio of dust to free molecules.

Hodge (1961) reported on giant particles 1.5 to 15μ in radius, collected on filter papers exposed on an airplane flying at 45,000 feet. For those particles larger than 3μ radius the mean concentration was of the order of 3×10^{-5} cm^{-3}. Most of these were found to be transparent or semitransparent, suggesting the possibility of terrestrial origin. Opaque particles of this size range numbered less than 10 per cent of the total. However, 54 per cent of all particles were in the $1.5 < r < 3\mu$ range, and they comprised 67.6 per cent of the opaque particles. Opaqueness might suggest meteoric origin, but Hodge concluded that the number of extraterrestrial particles was quite small.

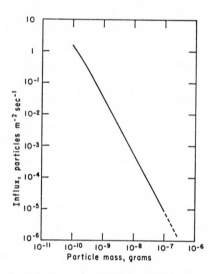

FIG. 4.11.—Average cumulative mass distribution from microphone systems on rockets and satellites as reported by Dubin and McCracken (1962).

Bigg (1961) has reported on examination of particles collected by filters on U-2 airplanes. Some of the particles gave the appearance of being of extraterrestrial origin, described as "opaque plates of metallic lustre," "opaque black shiny particles," "dull opaque particles," "translucent grains," "spheres, brightly colored or vitreous in appearance."

An ice core from the Greenland icecap representing approximately 39 years of snow accumulation has been melted, filtered to retention of half-micron sizes, and studied by Langway (1962). Black spherules having pitted surfaces have been found.

Particle Census-taking

In a census of particles in the atmosphere it is desirable to determine the size, number concentration, chemical composition, charge carried, if any, radioactivity, if any, at all points in the atmosphere. This is a task which may never be accomplished in the twentieth century. In fact, if one were asked to make a complete study of the atmospheric particles at a given place at a given time, one would be at a loss to find the apparatus and techniques to do so; the answer could only be approached in very small pieces with a variety of instruments and methods.

Following an approach taken by Lodge (1962), one might find it useful to list the various techniques in use. These may be divided into systems for (1) collecting, (2) sizing, (3) identifying, and (4) making general determinations, such as nucleating activity of aerosol-laden air or chemical analysis of rain and snow. The list below is believed to include the general principles underlying all measurement systems for this type of work.

Collection
    Gravitational sedimentation, as on dust-fall slides
    Washout: individual snowflake and raindrop residues
    Energy and inertia gradients
        Centrifugal impaction
        Linear impaction with or without inertia effects
    Electrostatic gradient
        Electrostatic precipitation
        Air-conductivity and ion-counting apparatus
    Thermal gradient
        Thermal precipitation
    Sifting, as with Millipore
Size determination
    Without collection
        Light scattering
        Elastic collision, as on microphone
        Photography of path in a force field
    After collection
        Microscope with growth factor
        Microscope without growth factor
Identification
    Analysis of total collection in bulk
        Chemical
        Physical (see below)
    Identification of individual particles
        Morphological, from atlas of micrographs
        Micrurgy, micromanipulation
        Hygroscopic behavior or volatility
        Magnetic properties
        Effects on gaseous agent
        Chemical reactions, as in spot tests
        Electrophoresis

Electron diffraction
X-ray diffraction
Special spectroscopic techniques
General determinations
 Nucleating activity
  Aitken counter
  Low-supersaturation counters
  Expansion chambers
  Cold-air mixing chambers
 Precipitation-water analysis
 Atmospheric optics, searchlight techniques, etc.

Measurement Systems

Details of apparatus and procedures for particle studies may be found in some of the references listed at the end of this chapter. The summary by Lodge (1962) containing an incomplete list of 108 references is recommended. Only the principles of some representative techniques will be discussed here.

Aitken counter.—In John Aitken's series of papers before the Royal Society of Edinburgh from 1883 to 1912 a number of versions of his particle counter were described, and since his time other scientific workers have introduced improvements. In all cases the same principle is involved. Air is drawn into a chamber by a graduated hand piston, which further serves to expand the chamber after the inlet is closed off. The condensed droplets are observed through an eyepiece in dark-field illumination from a mirror shining upward through a sheet of glass with small squares etched on it—a *graticule*. Droplets in one or more squares are counted and the number multiplied by the factor necessary to give the number per unit volume.

Pollak, Nolan, and O'Connor, beginning in 1946, published papers (1946, 1953, 1955, and 1961) on several improvements in Aitken counters, including a photoelectric counter. A counter based on these principles is marketed in the United States.

In the stratosphere balloon soundings shown in Figure 4.7, Junge used an automatic counter in which a motor operated the piston, the valves, and the controls of a stroboscopic flash tube and camera. The instrument was pressurized so as to insure droplets large enough to be photographed without strong magnification, and the temperature was maintained at 15 to 20° C. Water was injected into the chamber to make sure that the required supersaturations were achieved, since there would not be a high enough vapor pressure aloft to produce saturation at the high controlled temperatures.

For the Aitken counter, as for other expansion chambers, one can apply the relations of chapter 1 to obtain the saturation ratio in terms of the expansion ratio. In the expansion the specific humidity, assumed to be the saturated value at the beginning, is unchanged, so at any expanded state the saturation ratio is given by

$$S = \frac{w_{s1}}{w_{s2}} = \frac{p_{s1}/P_1}{p_{s2}/P_2} = \frac{p_{s1}P_2}{p_{s2}P_1},$$ (4.25)

where p_{s1} and p_{s2} are the saturation vapor pressures at the temperatures T_1, T_2 and P_1, P_2 are total pressures. From the adiabatic equation similar to (1.25) and (1.26) but in terms of pressure and volume, it is found that

$$\frac{P_2}{P_1} = \left(\frac{V_1}{V_2}\right)^{c_p/c_v},$$ (4.26)

so

$$S = \frac{p_{s1}}{p_{s2}} \left(\frac{V_1}{V_2}\right)^{c_p/c_v}.$$ (4.27)

The volume is read on the scale of the graduated piston, V_2/V_1 being the expansion ratio. The saturation vapor pressures, p_{s1} and p_{s2}, are read from tables as functions of temperature. Given the initial air temperature T_1, the adiabatically produced temperature T_2 is obtained from the adiabatic equation, and since

$$\frac{P_2}{P_1} = \left(\frac{T_2}{T_1}\right)^{m c_p/R} = \left(\frac{V_1}{V_2}\right)^{c_p/c_v},$$ (4.28)

then

$$T_2 = T_1 \left(\frac{V_1}{V_2}\right)^{R/m c_v}.$$ (4.29)

Cloud-nucleating activity.—To determine the cloud-nucleating activity, it is desirable to have an apparatus that will produce smaller supersaturations than in the Aitken counter. These conditions can be obtained with a properly controlled diffusion cloud chamber. In this chamber a temperature gradient is maintained in the air by cooling the bottom and warming the top. The top can be a warm, wet, porous plate or sponge for supplying warm water vapor, which diffuses downward. The coefficient of diffusion of water vapor in air and the coefficient of thermal diffusion stand in an essentially constant ratio, $D/\kappa \sim 1.2$, at all ordinary temperatures. Thus the distribution of water vapor in the chamber becomes a linear function of the temperature. This is not true of the saturation vapor pressure whose relationship to temperature is shown in the curve of Figure 2.2. By running a straight line between any two points on that curve, one cuts across a region of supersaturation which is smaller the less the temperature difference between the two end points. One reduces the supersaturation in the chamber by reducing the temperature difference between top and bottom. Condensation obviously occurs first in the part of the chamber having the greatest supersaturation. Wieland (1956) arranged a series of cells having different temperature gradients and thus obtained a spectrum of nucleating activity.

The problems involved in measuring nuclei of large and cloud size by means of thermal diffusion chambers have been discussed by Twomey (1963). He reports that the sample must be introduced into the chamber in such a way that it will not encounter warmer, moist surfaces or cooler, dry ones, since either can cause transient supersaturations which result in an unstable approach to equilibrium. Heating or cooling by conduction through the walls, even with small temperature gradients, can be serious if supersaturations of a few tenths of a per cent are sought. Furthermore, the vapor flux to the droplets may decrease with the increasing demand imposed by growing droplets. The shallower the chamber the smaller the last two effects become.

A "chemical diffusion" chamber has been devised by Twomey (1959) similar to a thermal diffusion chamber but with the vapor gradient maintained isothermally between a water-soaked pad at the top and a dilute aqueous solution of hydrochloric acid at the bottom. In the presence of HCl vapor in the chamber, the equilibrium water-vapor pressure is lowered because the condensed phase would at once include a certain amount of HCl. The water vapor, under diffusion at constant flux and constant air density, is established in a linear gradient. The equilibrium vapor pressure for the solution droplets is not linear because of the non-linear variation of the osmotic coefficient with concentration. Without going into detail it may be stated that the water vapor diffused from the source carries a higher vapor pressure than this equilibrium, so supersaturation with respect to any possible solution-droplet embryos is maintained.

With dilute solutions of HCl the supersaturations were low enough so that no droplets formed in filtered air. When unfiltered samples were introduced, a cloud would form. Supersaturations as low as 0.05 per cent were obtained. The clouds were photographed in a narrow beam of light and counts were made.

General collecting and sampling devices.—Nature's method of separating giant particles from smaller ones by fall in the field of gravity can be operated in the laboratory by causing the air to flow along a straight channel or tube. Particles will fall to the bottom of the tube at a distance from the entrance that depends on their size and initial height above the bottom. A cumulative size spectrum will be found along the tube such that nearly all sizes will be represented near the entrance, including some small ones that were already near the bottom when they entered. Near the exit only the smaller particles will be left. As a practical device, this crude apparatus is complicated by the requirement that in order to have purely gravitational fallout the flow must be laminar, and this type of flow is hard to produce.

Stronger accelerations than those of gravity can increase the collection rate, especially of the smaller particles. Centrifuge methods offer a solution. The acceleration can be made to vary if a conical centrifuge such as the "conifuge" of Sawyer and Walton (1950) is used.

In their "aerosol spectrometer" Goetz and Preining (1960) managed to keep

98

the flow rate relative to the confining walls small and laminar although the absolute velocity was high—6000 to 20,000 r.p.m. This condition was accomplished by passing the air through a helical channel around a rapidly spinning cone. The channel is a groove in the cone covered airtight by a flexible foil which can be removed for study of the particle deposit on it. The large particles are deposited near the entrance at the apex of the cone where the centrifugal action is just beginning, while the successively smaller particles are collected farther along the channel as larger radial velocities are achieved. The instrument is capable of collection and good size discrimination for a spectrum between 0.03 and 3.0µ diameter.

The so-called membrane filters actually are sieves which prevent particles above a certain size from passing through. Pore sizes of 0.5µ and less are found in commercially available membrane filters. Particles collected in this way can be studied individually by a variety of microscopic techniques.

Impactors are instruments usually dependent on the inertia of particles moving with the air in the vicinity of an obstacle. The particles strike the obstacle and can be made to cling to it while the air goes around it. An air filter such as one made of glass fibers can be regarded as a type of impactor containing a maze of obstacles. Theoretical impaction efficiencies on obstacles of various cross-sections in terms of particle size, air velocity, and air properties have been computed by Ranz and Wong (1952). For microscopic examination and size determination it is preferable to impact the particles on a series of flat surfaces. The cascade impactor of May (1945) has an air flow consisting of a series of right-angle turns against collecting plates at successively higher air speeds. The speeds are increased by drawing the air through smaller and smaller slits. Each successive plate collects smaller particles. The name comes from the fact that the air cascades downward over the plates toward a vacuum line at the bottom.

J. Dessens (1961) has devised a single-jet, single-plate impactor in which air moves downward toward a slide at an angle of 45 to 60 degrees, then along the slide under a slitlike restriction and into a larger volume. Particles of $r > 7\mu$ are nearly all deposited upstream from the slit, those of 1.8 to 4µ radius are mainly centered around the slit, while smaller ones are deposited downstream from it. There is considerable overlap in sizes in this distribution, but on either side of a 1.2-cm band around the slit there is discrimination between radii of 4µ and 1.8µ, with only 5 per cent of each on the "wrong" side.

In determining sizes of particles from inertial effects, it should be remembered that hygroscopic salts grow and increase their mass by taking on water even at moderate humidities. There is an advantage in making use of the hygroscopic property by adding water vapor and controlling the humidity as some investigators have done. For small particles supersaturation is desirable, as in the Aitken counter, and, as shown in later paragraphs, addition of moisture and other means of supersaturation are essential in studying ice nuclei.

99

Thermal precipitation is a process readily noted inside buildings in smoky cities, especially above radiators. Small particles collect on the cooler parts of the wall above a radiator. In some houses in severe climates the spacing of wood laths behind plaster on outside walls in a room is marked by dark lines where the plaster is in contact with the cold air within the walls and lighter colors along the areas in contact with the laths which insulate against the cold air. The theories of thermal precipitation are not uniformly accepted, but it is considered that the force carrying a particle down the temperature gradient is related to the transfer of momentum of gas molecules striking the particle. The stronger the temperature gradient and the lower the thermal conductivity of the particle, the greater will be the speed of precipitation. The speed also increases with decreasing size of the particle as long as the diameter is greater than the mean free path of the gas molecules—about 0.1μ in the lower atmosphere.

The simplest form of thermal precipitator consists of two parallel plates at different temperatures between which the air is drawn. Another method which, however, does not permit of metering the volume of air sampled, is to suspend a cold surface in the ambient air.

Electrostatic precipitation is a common method of cleaning air and can also be used for aerosol sampling. The particles are given a charge as the air passes around an ionizing wire or ring at the entrance to the duct. An electric field is imposed between the walls of the tubular duct and a rod serving as an electrode mounted in the center and carrying charge of the same sign as the particles. The particles migrate to the walls at a speed depending on their mobility as ions, which is a function of size, and on the strength of the field. Small particles are precipitated faster than large ones because of their higher mobility.

Optical methods, especially those based on scattering, are used to determine the sizes and number concentrations in volumes of air. The measurements of twilight and of sky radiance at various altitudes have been mentioned in an earlier part of this chapter as a means of determining particle distributions. The scattering of sunlight, such as that producing sky brightness or twilight, depends on scattering which is a function of particle size in relation to wavelength and, of course, of the number concentration.

Identification of particles.—Analytical chemical methods may be used for particles of micron size using modified spot tests, as shown by Seely (1952, 1955), Fidele and Vittori (1953), Vittori (1955), and developed to a high degree by Lodge *et al.* (1954, 1956, 1959, 1960a, b) and Tufts (1958, 1959, 1960). When reacted upon by a suitable reagent, an inorganic particle deposits the reaction product in a circle or "halo" on the substrate around it. The reagent used reacts on ions of only one substance, e.g., mercurous fluosilicate on chloride ion, ammonium ferrocyanide on calcium ion, lead nitrate on sulfate ion, etc. The size of the halo is related to the particle size. The particles are impacted on a film of gelatin containing the desired reagent or, better still, on a membrane filter which can subsequently be cut into pieces and each piece floated on the surface of a different reagent to test for various substances in the same sample. If the air is metered, one can obtain the number concentration in addition to the size and composition of the particles. The technique is particularly useful for monitoring the giant nuclei. The data on the distributions of giant chloride and sulfate particles given earlier in this chapter were obtained by this method. With certain special techniques, particles in the submicron range can be studied.

Referring back to Figure 2.4 in chapter 2, one notes that for the substance NaCl used in the example, the dry particle begins to form a saturated solution at a relative humidity of about 75 per cent. The humidity at which this occurs at a given temperature varies with the substance, thus providing discrimination with no overlap in values for the simple compounds likely to occur in the atmosphere.[5] Twomey (1953) devised equipment and performed this type of test. The particles were collected at the ground and in an airplane on spiders' threads (available commercially) stretched on a frame. After exposure to air moving at 60 to 75 m sec^{-1}, the frame was placed on a microscope stage and air from a controllable humidifier was introduced. The humidity was gradually increased until the phase transition was seen occurring on the various particles, which were then identified through the humidity reading. Hygroscopic particles of mass down to about 10^{-12} g could be identified in this way.

Freezing Nuclei

It has long been realized that the characteristic cloud condensation nuclei do not serve as nuclei for freezing or crystallization. The natural freezing nuclei appear to be mainly insoluble substances. They are active only at large supersaturation (subcooling).

Surveys of these nuclei in the atmosphere have been based mainly on measures of nucleating activity. Several designs of cloud chamber aimed at determining the number of ice crystals formed per unit volume as a function of temperature in the chamber have been used. Three different principles of operation have been adopted: diffusion chambers, mixing chambers, and expansion chambers.

Diffusion chambers are not well suited for this purpose because the large subcooling needed for nucleation demands a temperature gradient too large to permit easy temperature determination at the nucleating site. Schaefer (1949) has successfully photographed ice-crystal layers in a freezing box opened at the top, showing the crystals growing as they sank to lower, colder parts of the box. An open box of this type has the disadvantage that air from the immediate environment may contaminate the sample.

[5] Lagord (1961) determined that for common hygroscopic salts the critical humidities vary little with temperature down to temperatures considerably below 0° C.

In the mixing cloud chamber a volume of atmospheric air is stirred inside a cold chamber along with a cloud of droplets from distilled water heated in a flask. The chamber is cooled while the air is continually stirred, and the temperature is noted when the ice crystals begin to be visible in a beam of light. A box of this type designed by Smith and Heffernan (1954) has been used in airplane flights. In flights reported by Smith *et al.* (1956) in Australia and Arizona, nucleating activity began at temperatures ranging from -17 to $-36°$ C. The day-to-day variation was much larger in Australia than in Arizona, possibly because both sets of flights were in January, when it was summer in Australia and winter in Arizona.

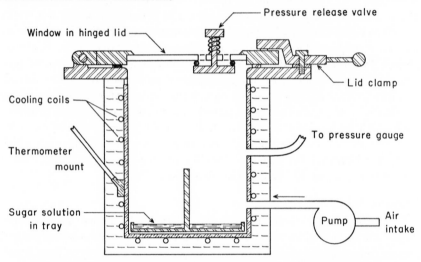

Fig. 4.12.—Schematic design of U.S. Weather Bureau version of Bigg-Warner box

A continuously recording mixing cloud chamber has been developed by Lutterman *et al.* (1959) which counts scintillations of ice crystals formed in a subcooled fog running through the chamber.

A chamber based on a cooling design by Bigg (1957) and an expansion-type counter by Warner (1957) has been used especially, but not exclusively, for determinations at the ground. Bigg got around the subjectivity of counting crystals in the cloud volume by allowing the crystals to grow and fall into a tray of saturated sugar solution subcooled to $-12°$ C. In the sugar solution the ice crystals grow rapidly, a millimeter per minute or more. Against the background of the black-painted tray the crystals are easily and quickly counted.

The Warner box, modified by the U.S. Weather Bureau, is shown in Figure 4.12. It consists of a mechanical refrigerating system with cooling coils around the cloud chamber. The sugar-solution tray is placed in the bottom, the glass-windowed lid is clamped on, and air is pumped in to produce a slight overpressure while cooling proceeds. Then the pressure-relief valve is made to spring open, allowing a sudden expansion to occur. For operations at low altitudes it is not necessary to add water vapor. The temperature is controllable to about −30° C.

In this box, as well as all other boxes, great care is necessary to avoid the formation of ice crystals on the walls which can serve to nucleate new crystals in the chamber. The walls are frequently coated with glycerol to prevent this formation.

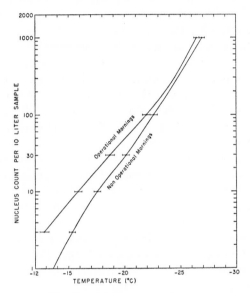

Fig. 4.13.—Bourquard's curves from two summers of daily measurements of ice-nucleating activity at the University of Chicago radar field site near West Plains, Missouri. "Operational mornings" were on days selected as favorable for radar and airborne studies of convective clouds.

Results of measurements with the Weather Bureau box taken at the ground in a sparsely populated area near West Plains, Missouri, by Bourquard (1963) are shown in Figure 4.13. The data are averaged from daytime collections of three summers. They show the expected relation between nucleating activity and temperature. Wide variations not readily explainable from meteorological or other conditions are smoothed in the averages. Higher average counts in the mornings than in the afternoons are found, possibly explainable as a result of accumulation of nuclei of ground origin under the nocturnal temperature inversion.

Bigg and Meade (1959) have developed a continuous counter wherein the sugar solution is coated on a traveling belt and continuous air injection is provided. An elaboration of this device has been accomplished by Admirat (1963).

103

Bigg (1960) compared results of measurements of ice-nucleus concentrations taken between 1955 and 1960 at a number of places in both northern and southern hemispheres. It was found that mixing chambers in use at that time gave consistently higher values of the nucleating activity at $-20°$ C than did the expansion chambers. It was also noted that values obtained in the northern hemisphere, especially near populated areas, were higher than in the southern hemisphere, a situation ascribed to smoke pollutants, as previously demonstrated by Soulage (1958). Admirat (1962) has examined ice nuclei of industrial origin and has found that the smokes contain many active nuclei but their activity diminishes rapidly, apparently because of the effects of inactive particles or gases in "poisoning" or masking their surfaces.

Lodge and Bravo (1962) have developed an apparatus for collecting particles by drawing air through a membrane filter, placing it on a cold block, closing it off, supplying moisture to it, and then pouring sugar solution onto the filter surface. Crystals form in the sugar solution at sites containing suitable nuclei. Bigg *et al.* (1962) report similar methods and note important effects of humidification time and volume of air filtered.

Ice-crystal nuclei.—The best information concerning the substances which serve as natural ice nuclei in the atmosphere has been obtained by Kumai (1961, 1962) and Isono (1955). Snow crystals precipitating in unpolluted atmospheres —parts of Japan, Lake Superior, Greenland—are collected on collodion-coated slides of the electron microscope. To prevent the snow crystals from subliming too rapidly, the collection is made in an igloo with a hole in the roof or, in the case of Greenland, in the entrance to a snow tunnel. During sublimation each crystal is photographed at short time intervals under an optical microscope until the last photograph shows only where the center was. This spot, located with reference to the grid on the slide, is later examined and photographed under the electron microscope. At what was the center of each crystal a particle is found which is considered to be the nucleus. Aerosols located at other points also are found, but they are thought to have been captured by the snow crystal either within the mother cloud or during fall beneath the cloud. In the electron microscope the substance of the nucleus is identified by electron diffraction or, in some cases, by its appearance.

Of 271 snow crystals from Houghton, Michigan studied in detail, Kumai found that 87 per cent had clay mineral particles as their nucleus. Of 356 snow crystals collected on the elevated icecap of northern Greenland in summer, 85 per cent had clay mineral particles at their centers. More than half the clay minerals were identified as kaolin, the remainder being montmorillonite, illite, attapulgite, and related groups. Of the total number of center particles, only four—two in each geographic location—were hygroscopic substances. A few combustion particles were found in the Houghton, Michigan, series. About 10 per cent of the nuclei from both places could not be identified by the avail-

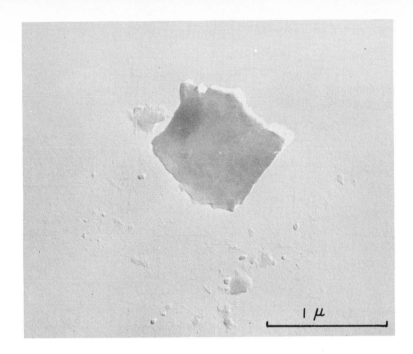

PLATE 4.1.—Example of electron-microscope study of snow-crystal nucleus by Kumai (1961): The upper photograph is an electron micrograph of particle remaining at center of snow crystal after sublimation; below is an electron-diffraction pattern of the same particle. The particle was identified as kaolin.

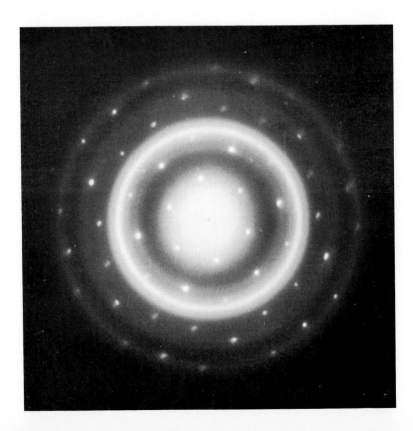

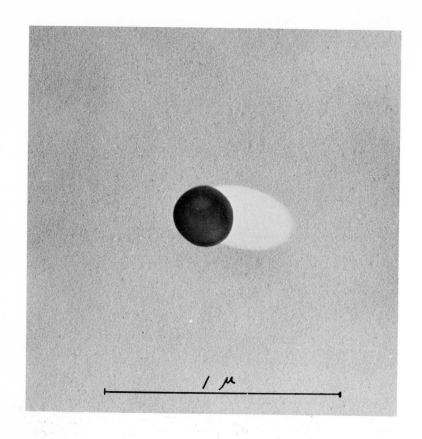

PLATE 4.2.—Electron micrograph by Kumai (Kumai and Francis, 1962) of unidentified spherical particle-forming nucleus of snow crystal collected on the Greenland ice cap.

able techniques. In 13 of the Greenland snow crystals no center nucleus could be observed, suggesting the possibility of homogeneous nucleation.

An example of the identification is represented in the two photographs in Plate 4.1, with the hexagonal diffraction pattern of the nucleus shown in *b*. The lattice values deduced from the Debye-Sherrer pattern coincide with those for kaolin.

Kumai's results might be interpreted as failing to verify a hypothesis, founded on indirect evidence, that meteoric particles serve as ice nuclei (Bowen, 1953, 1956). In the Australian stratosphere flights reported by Bigg (1961) it was found that the nucleating activity at −15° C in the stratosphere was greater than at the surface. Some of Kumai's unidentified particles, such as that shown in Plate 4.2, suggest meteoric origin.

REFERENCES

ADMIRAT, P. (1962), *Bull. Obs. Puy de Dôme*, No. 2, p. 87.

———. (1963), *J. Recherches Atmos.*, 1, 1.

AITKEN, J. (1884 *et seq.*); see *Collected Scientific Papers*, ed. C. G. KNOTT (Cambridge: Cambridge University Press, 1923).

ALEXANDER, W. M., McCRACKEN, C. W., SECRETAN, L., and BERG, O. E. (1962), *Trans. Amer. Geophys. U.*, 43, 351.

BEST, A. C. (1950), *Quart. J. Roy. Met. Soc.*, 76, 16.

BIGG, E. K. (1957), *Tellus*, 9, 394.

———. (1960), *Bull. Obs. Puy de Dôme*, No. 3, 89.

———. (1961), *Abstracts, Internat. Conf. Cloud Physics*, Canberra and Sydney, Australia.

BIGG, E. K., and MEADE, R. T. (1959), *Bull. Obs. Puy de Dôme*, No. 4, p. 125.

BIGG, E. K., MOSSOP, S. C., THORNDIKE, N. S. C., and MEADE, R. T. (1962), *Bull. Obs. Puy de Dôme*, No. 3, p. 141.

BLANCHARD, D. C., and WOODCOCK, A. H. (1957), *Tellus*, 9, 145.

BOURQUARD, D. (1963), *J. Atmos. Sci.*, 20, 386.

BOWEN, E. G. (1953), *Australian J. Phys.*, 6, 490.

———. (1956), *J. Met.*, 13, 142.

BUDDHUE, J. D. (1950), *Meteoritic Dust* (Albuquerque: University of New Mexico Press).

BYERS, H. R. (1957), *Quart. J. Roy. Met. Soc.*, 83, 387.

BYERS, H. R., SIEVERS, J. R., and TUFTS, B. J. (1957), in *Artificial Stimulation of Rain*, ed. H. WEICKMANN and W. SMITH (New York: Pergamon Press), p. 47.

CARLETON, N. P. (1962), *J. Atmos. Sci.*, **19**, 424.

CHAGNON, C. W., and JUNGE, C. E. (1961), *J. Met.*, **18**, 746.

CONWAY, E. J. (1943), *Proc. Roy. Irish Acad.*, **48**, 119 and 161.

COSTE, J. H., and WRIGHT, H. L. (1935), *Phil. Mag.*, **20**, 209.

CUNNINGHAM, E. (1910), *Proc. Roy. Soc. London, Ser. A*, **83**, 357.

DAVIES, C. N. (1945), *Proc. Phys. Soc., Ser. A*, **57**, 259.

DESSENS, J. (1961), *Bull. Obs. Puy de Dôme*, No. 1, p. 1.

DUBIN, M., and MCCRACKEN, C. W. (1962), *Astron. J.*, **67**, 248.

DURBIN, W. G., and WHITE, G. D. (1961) *Tellus*, **13**, 260.

EINSTEIN, A. (1905), *Ann. Physik* (Leipzig), **17**, 549.

ERIKSSON, E. (1959), *Tellus*, **11**, 375.

———. (1960), *ibid.*, **12**, 63.

FEDELE, D., and VITTORI, O. (1953), *Riv. Met. Aeron.*, **13**, No. 4, p. 9.

FENN, R. W. (1960), *J. Geophys. Res.*, **65**, 3371.

GAMBELL, A. W., JR. (1962), *Tellus*, **14**, 91.

GERHARD, E. R., and JOHNSTONE, H. F. (1955), *Indus. Eng. Chem.*, **47**, 972.

GOETZ, A., and PREINING, O. (1960), in *Physics of Precipitation*, ed. H. WEICK-MANN (Amer. Geophys. U. Monogr. No. 5), p. 164.

GREENFIELD, S. M. (1957), *J. Met.*, **14**, 115.

HAAGEN-SMIT, A. J. (1956), *Indus. Eng. Chem.*, **48**, 65a.

HAYAMA, S., and TOBA, Y. (1958), *J. Oceanogr. Soc. Japan*, **14**, 145.

HODGE, P. W. (1961), *Smithsonian Contrib. Astrophys.*, **5**, No. 10.

ISONO, K. (1955), *J. Met.*, **12**, 456.

———. (1959), *Japanese J. Geophys.*, **20**, 1.

JUNGE, C. E. (1955), *J. Met.*, **12**, 13.

———. (1957), in *Artificial Stimulation of Rain*, ed. H. WEICKMANN and W. SMITH (New York: Pergamon Press), p. 3.

———. (1958), in *Advances in Geophysics*, ed. H. LANDSBERG and J. VAN MIEGHEM (New York: Academic Press), **4**, 1.

———. (1960), *J. Geophys. Res.*, **65**, 227.

———. (1961), *J. Met.* **18**, 501.

JUNGE, C. E., CHAGNON, C. W., and MANSON, J. E. (1961), *J. Met.*, **18**, 81.

JUNGE, C. E., and MANSON, J. E. (1961), *J. Geophys. Res.*, **66**, 2163.

JUNGE, C. E., and WERBY, R. T. (1958), *J. Met.*, **15**, 417.

KNELMAN, F., DOMBROWSKI, N., and NEWITT, D. M. (1954), *Nature*, **173**, 261.

KUMAI, M. (1961), *J. Met.*, **18**, 139.

KUMAI, M., and FRANCIS, K. E. (1962), *J. Atmos. Sci.*, **19**, 474.

LAGORD, J. (1961), *Bull. Obs. Puy de Dôme*, No. 2, p. 87.

LANGWAY, C. C., JR. (1962), *Internat. Assn. Sci. Hydrol., Commission Snow and Ice*. Pub. No. 58, p. 101.

LODGE, J. P., JR. (1954), *Anal. Chem.*, **26**, 1829.

———. (1955), *J. Met.*, **12**, 493.

————. (1959), *Nubila*, **2**, 58.

————. (1960), in *Physics of Precipitation*, ed. H. WEICKMANN (Amer. Geophys. U. Monogr. No. 5), p. 252.

————. (1962), in *Advances in Geophysics*, ed. H. LANDSBERG and J. VAN MIEGHEM (New York: Academic Press), **9**, 97.

LODGE, J. P., JR., and BRAVO, H. (1962), *Bull. Obs. Puy de Dôme*, No. 10, p. 81.

LODGE, J. P., JR., FERGUSON, J., and HAVLIK, B. R. (1960a), *Anal. Chem.*, **32**, 1206.

LODGE, J. P., JR., McDONALD, A. J., JR., and VIHMAN, E. (1960b), *Tellus*, **12**, 184.

LODGE, J. P., JR., ROSS, H. F., SUMIDA, W. K., and TUFTS, B. J. (1956), *Anal. Chem.*, **28**, 423.

LODGE, J. P., JR., and TUFTS, B. J. (1956), *Tellus*, **8**, 184.

LUTTERMOSER, R. L., BOUTIN, L. J., and HAWLEY, W. G. (1959), Status Rep. No. 3 to National Science Foundation, Borg Warner Corp., Anaheim, Calif.

McCRACKEN, C. W., ALEXANDER, W. M., and DUBIN, M. (1961), *Nature*, **192**, 441.

McDONALD, J. E. (1962), *Science*, **135**, 435.

MASON, B. J. (1955), *Nature*, **174**, 470.

MAY, K. R. (1945), *J. Sci. Instr.*, **22**, 187.

MILLIKAN, R. A. (1923), *Phys. Rev.*, **22**, 1.

MOORE, D. J., and MASON, B. J. (1954), *Quart. J. Roy. Met. Soc.*, **78**, 596.

NEWKIRK, G., JR., and EDDY, J. A. (1964), *J. Atmos. Sci.*, **21**, 35.

NOLAN, P. J., and POLLAK, L. W. (1946), *Proc. Roy. Irish Acad., Ser. A*, **51**, 9.

O'CONNOR, T. C. (1961), *Quart. J. Roy. Met. Soc.*, **87**, 105.

O'CONNOR, T. C., SHARKEY, W. P., and O'BROLCHAIN, C. (1959), *Geofys. Pura e Appl.*, **42**, 109.

OKITA, T. (1962), *J. Met. Soc. Japan*, **40**, 163.

ÖPIK, E. J. (1956), *Irish Astron. J.*, **4**, 84.

POLLAK, L. W. (1952), *Geofys. Pura e Appl.*, **22**, 75.

POLLAK, L. W., and MURPHY, T. (1953), *Geofys. Pura e Appl.*, **25**, 44.

POLLAK, L. W., and O'CONNOR, T. C. (1955), *Geofys. Pura e Appl.*, **32**, 139.

RANZ, W. E., and WONG, J. B. (1952), *Indus. Eng. Chem.*, **44**, 1371.

RIEHL, H. (1954), *Tropical Meteorology* (New York: McGraw-Hill Book Co.), p. 53.

SAGALYN, R. C., and FAUCHER, G. A. (1956), *Quart. J. Roy. Met. Soc.*, **82**, 428.

SAWYER, K. F., and WALTON, W. H. (1950), *J. Sci. Instr.*, **27**, 272.

SCHAEFER, V. J. (1949), *Chem. Rev.*, **44**, 291.

SEELY, B. K. (1952), *Anal. Chem.*, **24**, 576.

————. (1955), *ibid.*, **27**, 93.

SINGLETON, F., and DURBIN, W. G. (1962), *Quart. J. Roy. Met. Soc.*, **88**, 315.

SMITH, E. J., and HEFFERNAN, K. J. (1954), *Quart. J. Roy. Met. Soc.*, **80**, 182.
SMITH, E. J., KASSANDER, A. R., JR., and TWOMEY, S. (1956), *Nature*, **177**, 82.
SMOLUCHOWSKI, M. (1918), *Zeitschr. Phys. Chem.*, **92**, 129.
———. (1926), *Kolloid-Chem. Beiheft*, **22**, 192.
SOULAGE, G. (1958), *Bull. Obs. Puy de Dôme*, No. 3, 89.
STOKES, G. G. (1850), *Trans. Cambridge Phil. Soc.*, **9**, 8. See also STOKES, G. G., *Mathematical and Physical Papers* (Cambridge: Cambridge University Press, 1941), **3**, 58.
TOBA, Y. (1959), *J. Oceanogr. Soc. Japan*, **15**, 121.
———. (1961), *Mem. College Sci. Univ. Kyoto, Ser. A*, **29**, 313.
TUFTS, B. J. (1959), *Anal. Chem.*, **31**, 238.
———. (1960), *Anal. Chim. Acta*, **23**, 209.
TUFTS, B. J., and LODGE, J. P. (1958), *Anal. Chem.*, **30**, 300.
TUNITZKI, N. (1946), *Phys. Chem.* (USSR), **20**, 1137.
TWOMEY, S. (1953), *J. Appl. Phys.*, **24**, 1099.
———. (1959), *Geofys. Pura e Appl.*, **43**, 227.
———. (1960), *Bull. Obs. Puy de Dôme*, No. 1, p. 1.
———. (1963), *J. Recherches Atmos.*, **1**, 101.
VERZÁR, F., and EVANS, H. D. (1959), *Geofys. Pura e Appl.*, **43**, 259.
VITTORI, O. (1955), *Geofys. Pura e Appl.*, **31**, 90.
VOLZ, F. E., and GOODY, R. M. (1962), *J. Atmos. Sci.*, **19**, 385.
WARNER, J. (1957), *Bull. Obs. Puy de Dôme*, No. 1, p. 33.
WEICKMANN, H. (1957), in *Artificial Stimulation of Rain*, ed. H. WEICKMANN and W. SMITH (New York: Pergamon Press), p. 81.
WENT, F. W. (1960), *Proc. Nat'l. Acad. Sci.*, **48**, 309.
WHYTLAW-GRAY, R., and PATTERSON, H. S. (1932), *Smoke* (London: Arnold and Co.).
WIELAND, W. (1956), *Zeitschr. Angew. Math. Phys.*, **7**, 428.
WIGAND, A. (1919), *Ann. Phys.* (Leipzig), **59**, 689.
WOODCOCK, A. H. (1953), *J. Met.*, **10**, 362.
WOODCOCK, A. H., KIENTZLER, C. F., ARONS, A. B., and BLANCHARD, D. C. (1953), *Nature*, **172**, 1145.

5

THE INITIAL
GROWTH OF DROPLETS
AND ICE CRYSTALS
IN CLOUDS

After a droplet or ice crystal has been nucleated and has surpassed the free-energy barrier or critical radius, it enters a stage of growth by diffusion of vapor to it. This growth is maintained as long as the saturation ratio or super-saturation required for it prevails. A third stage of growth caused by collision and coalescence with other drops or crystals will be treated in chapter 6.

The growth by diffusion is extremely rapid at first, the droplet or crystal growing from the size of the nucleus to visible size in a fraction of a second, but soon slows down. The reverse process, evaporation or sublimation, is governed by the same laws of diffusion and will be treated in this .chapter along with growth.

Theory of Droplet Growth and Evaporation

The flux of vapor from a droplet (evaporation) is

$$\frac{1}{A}\frac{dM}{dt} = -D\frac{d\rho_w}{dr},\qquad(5.1)$$

where D is the diffusivity of water vapor in air, A is the surface area of the droplet, and M is the mass of water involved in the exchange. The gain in mass by diffusion-condensation on a droplet is therefore

$$\frac{dM}{dt} = 4\pi r^2 D\frac{d\rho_w}{dr}.\qquad(5.2)$$

A droplet in a vapor environment presents at its surface a discontinuity between the two phases. It has a vapor density at its surface corresponding to its vapor tension, while a uniform vapor density prevails in the ambient vapor. This distribution requires replacing the continuous gradient $d\rho_w/dr$ by the value $\rho_{w(r)}$ for the vapor density at the surface of the droplet at radius r, and by ρ_w at $r = \infty$ for the ambient vapor density.[1] The growth rate in terms of radius, dr/dt, of the droplet under these conditions can be obtained by integrating (5.2) for a given flux, dM/dt, then inserting the inherent dependence of dM/dt on dr/dt. We rearrange (5.2) and form the integral

$$\frac{dM}{dt} \int_r^\infty \frac{dr}{r^2} = 4\pi D \int_{\rho_{w(r)}}^{\rho_w} d\rho_w , \qquad (5.3)$$

resulting in

$$\frac{dM}{dt} = 4\pi D r [\rho_w - \rho_{w(r)}] = \rho_L \frac{dV}{dt} = \rho_L 4\pi r^2 \frac{dr}{dt} , \qquad (5.4)$$

so

$$r \frac{dr}{dt} = \frac{D}{\rho_L} [\rho_w - \rho_{w(r)}] . \qquad (5.5)$$

By substitution from the equation of state, the relation can be expressed in terms of vapor pressure as

$$r \frac{dr}{dt} = \frac{D m_w}{\rho_L R T} (p - p_r) , \qquad (5.6)$$

assuming that the vapor at the droplet is at the same temperature as in the ambient air.

During a balanced state, the latent heat added must be liberated to the environment, and vice versa. The rate of release of latent heat is, from (5.4),

$$L \frac{dM}{dt} = 4\pi r L D [\rho_w - \rho_{w(r)}] \qquad (5.7)$$

and the diffusion of heat away from the droplet is, as in the case of vapor diffusion, proportional to the thermal diffusivity κ multiplied by the difference in heat content at constant pressure between the droplet and the ambient air,

$$\frac{dQ}{dt} = 4\pi r \kappa \rho c_p (T_r - T) , \qquad (5.8)$$

or, in terms of the more commonly used thermal conductivity, $K = \kappa \rho c_p$,

$$\frac{dQ}{dt} = 4\pi r K (T_r - T) . \qquad (5.9)$$

---

[1] Infinity may be considered as lying between the droplet and its neighbors, which would be some hundreds of droplet radii away in dense clouds, farther in attenuated clouds.

For balance, $dQ/dt = L(dM/dt)$ and, substituting dM/dt from (5.7), we have

$$\frac{\rho_w - \rho_{w(r)}}{T_r - T} = \frac{K}{DL} \qquad (5.10)$$

or

$$\frac{p - p_r}{T(T_r - T)} = \frac{RK}{m_w DL}. \qquad (5.11)$$

Note that this equilibrium is essentially the same as that over a wet bulb, derived in chapter 1, equation (1.39). When the ambient vapor is at supersaturation with respect to the droplet, $p > p_r$, vapor flows to the droplet, and heat flows away for $T_r > T$. Equation (5.6) is substituted in equation (5.11) giving

$$r\frac{dr}{dt} = \frac{K}{\rho_L L}(T_r - T). \qquad (5.12)$$

We desire to combine the vapor and heat flux effects to obtain the growth rate in terms of the observable saturation ratio of the droplet environment. As the droplet grows beyond its critical radius the curvature and solution effects quickly disappear. The vapor tension of the droplet is the same as that of a plane surface of pure water at the temperature T_r of the droplet. We consider three vapor pressures as follows:

p, actual ambient vapor pressure;

$p_0(T)$, saturation vapor pressure at ambient temperature, T;

$p_0(T_r)$, vapor pressure over a droplet, considered as a plane, pure surface at temperature T_r.

Equation (5.6) for growth without consideration of the heat effect is written in these terms as

$$r\frac{dr}{dt} = \frac{D m_w}{\rho_L RT}[p - p_0(T_r)], \qquad (5.13)$$

which, divided by $p_0(T)$, becomes

$$\frac{p - p_0(T_r)}{p_0(T)} = \frac{\rho_L RT}{D m_w p_0(T)} r\frac{dr}{dt}. \qquad (5.14)$$

Mason (1957) developed the steps of combining this expression with (5.12) to obtain a growth equation which includes both the vapor and heat diffusion effects and which depends only on the saturation ratio and temperature of the ambient air.

The first step is to integrate the Clapeyron-Clausius equation for an ideal vapor as in (2.34) between the limits $p_0(T)$ to $p_0(T_r)$ and T to T_r, with the result

$$\ln\frac{p_0(T_r)}{p_0(T)} = \frac{m_w L}{R}\left(\frac{1}{T} - \frac{1}{T_r}\right) = \frac{m_w L}{RT_r T}(T_r - T). \qquad (5.15)$$

111

For the small temperature differences with which we are dealing around the individual droplets, we may substitute T^2 for $T_r T$ in this equation and write

$$\ln \frac{p_0(T_r)}{p_0(T)} = \frac{m_w L}{RT^2}(T_r - T).$$
(5.16)

The next step is to substitute for $T_r - T$ from equation (5.12) to obtain

$$\ln \frac{p_0(T_r)}{p_0(T)} = \frac{m_w L^2 \rho_L}{KRT^2} r \frac{dr}{dt}.$$
(5.17)

Finally a substitution is made into (5.16) of this last expression for $p_0(T_r)/p_0(T)$ in exponential form, a substitution which is best seen after writing (5.14) in the form

$$\frac{p}{p_0(T)} = \frac{p_0(T_r)}{p_0(T)} + \frac{\rho_L RT}{p_0(T) D m_w} r \frac{dr}{dt},$$
(5.18)

the substitution from equation (5.17) being for the first term on the right. The result is

$$\frac{p}{p_0(T)} = \exp\left(\frac{m_w L^2 \rho_L}{KRT^2} r \frac{dr}{dt}\right) + \frac{\rho_L RT}{p_0(T) D m_w} r \frac{dr}{dt}.$$
(5.19)

The relation is now expressed in terms of the ambient saturation ratio S. In symbolic form the expression is

$$S = e^{ax} + bx,$$
(5.20)

where $x \equiv r(dr/dt)$. When $ax \ll 1$ as in this case, an expansion of e^{ax} shows that it may be simplified to $1 + ax$. Thus we have

$$S = 1 + ax + bx,$$
(5.21)

and solving for x we obtain the final form of the growth equation as

$$r \frac{dr}{dt} = \frac{S-1}{a+b} = \frac{S-1}{(m_w L^2 \rho_L / KRT^2) + (\rho_L RT / p_0(T) D m_w)}.$$
(5.22)

The numerator is the supersaturation. In the case of evaporation it would be negative (subsaturation). In the denominator, L, ρ_L, K, and D vary slightly with the temperature, the last also with the total pressure, so the two temperature terms and the ambient saturation vapor pressure, which is quite sensitive to temperature, are the important variables of the denominator.

In many cloud situations the various terms would change with the decreasing temperature and total pressure in the saturation-adiabatic ascent of the air. In some situations, such as the evaporation of a drop in the laboratory, the ambient terms, including S, would remain constant and the integration of equation (5.22) would result simply in

$$r^2 = r_0^2 + \frac{2(S-1)}{a+b}(t - t_0).$$
(5.23)

For the real atmospheric situations it is best to program the integration for an electronic computer.

For a given temperature, pressure, and supersaturation, the growth rate is faster the smaller the droplet, but a small droplet does not reach the size of an initially larger one. In terms of mass the buildup is faster the larger the droplet.

To express the growth in terms of mass, equation (5.22) may be written with the aid of equation (5.4) as

$$\frac{dM}{dt} = \rho_L 4\pi r \left(r \frac{dr}{dt} \right) = \frac{4\pi r (S-1)}{a'+b'}, \qquad (5.24)$$

where $a' + b' = (a + b)/\rho_L$. This eliminates ρ_L from consideration since $a + b$ contains a factor ρ_L.

Ventilation Effect and Evaporation

In applying the equation (5.22) to the evaporation of drops, a ventilation factor must be included. Frössling (1938) showed that the air flow past a drop increased the evaporation by a factor dependent on the square root of the Reynolds number. Kinzer and Gunn (1951) carried out a series of experiments in which terminal velocity as a function of mass was first determined (Gunn and Kinzer, 1949), then, in the evaporating case, mass changes were determined from the changes in terminal velocity. The times required to travel fixed distances in free fall in a vertical tube were determined by two methods—photographically, and by having charged drops fall through a series of induction rings to produce an electronically recorded pulse at each ring. They found that the ventilation factor could best be represented by $1 + 0.22\, FRe^{1/2}$, where F is itself a function of the Reynolds number Re. Over the range of Reynolds numbers for falling raindrops of radius greater than 0.25 mm ($Re > 100$), F is very close to one, but for smaller sizes ($Re < 100$) it varies markedly with Reynolds number, rising to a peak of about 2.2 at Re of about 2.1 ($r \sim 60\mu$), then dropping off to zero at $Re < 1.0$ ($r < 40\mu$). In computing the growth of drops in clouds the ventilation effect is not used because the growth by diffusion is negligible compared with that by coalescence at sizes greater than about 40μ radius, and for smaller sizes $F \cong 0$.

The evaporation of raindrops as they fall from a cloud may be computed by including the ventilation factor in equation (5.22) or, in terms of height, $dr/dz = (dr/dt)(dt/dz)$, by

$$\frac{dr}{dz} = \frac{(S-1)(1+FRe^{1/2})}{r(u_T - U)(a+b)}, \qquad (5.25)$$

where u_T is the terminal velocity and U is the draft velocity, positive for updraft. Integrated between the ground and the cloud base at height h, the expression becomes

$$\int_{r_0}^{r_h} \frac{u_T - U}{1 + FRe^{1/2}}\, r\, dr = \int_0^h \frac{S-1}{a+b}\, dz. \qquad (5.26)$$

113

The terms on the left are functions of radius and those on the right depend, through temperature, pressure, and humidity, upon the height. But the terms in r change at a rate dependent on the humidity, so all variables change if S is not constant. Mason (1952) integrated the expression by taking 10μ intervals of r starting with $r_0 = 100\mu$ at the ground and with various mean relative humidities of the subcloud air, with S constant with height. He then determined r_h for various values of cloud height, h. For example, for a cloud base at 1000 m, a drop falling through air of 90 per cent relative humidity with an updraft U of 10 cm sec^{-1} would leave the cloud with a radius of 220μ in order to reach the ground with a radius of 100μ. For $h = 250$ m and 98 per cent relative humidity, r_h would be 117μ if $r_0 = 100\mu$.

Growth of a Population of Droplets

We have seen in chapter 4 that a broad spectrum of sizes of condensation nuclei is available in the atmosphere. For a given supersaturation all droplets larger than the critical size can grow in the diffusion-growth stage. The increase in size is faster the smaller the droplet, as shown, for instance, by (5.24), such that $dr/dt \propto r^{-2} dM/dt \propto r^{-1}(S - 1)$. Thus, as long as the vapor supply lasts, the small ones approach the large ones in size and the spectrum narrows. It is interesting to compute the distribution of sizes from given distributions of nuclei and the various supersaturations resulting from various rates of cooling and utilization of vapor.

In 1949 Howell published results of computations of size spectra resulting from diffusion growth in clouds. Later results by Mordy (1959) and Neiburger and Chien (1960), using electronic computers, added interesting information concerning the growth processes. Machine computation is called for because a number of quantities are changing together, with some feedbacks between them.

The variables involved include the size spectrum and water activity of the various nuclei, the changing droplet radii, the time variable, and the degree of supersaturation. The latter depends on the rate of ascent and also is interdependent with the other variables. In solving the problem the usual procedure is to take separate models of given species and distributions of nuclei and given ascent rates. Starting with each nuclear size, one computes a number of growth curves, each droplet size class ending up with the same number concentration as that of the nucleus size class from which it was drawn. A series of such computations starting in various parts of the nuclei spectrum is made while the droplets interact with the supersaturation. The ascensional cooling increases the supersaturation at first, but it soon is decreased as water vapor is used in condensation. In the larger sizes, sedimentation adds a complication.

Two examples of growth curves computed, in one case, by Mordy and, in the other, by Neiburger and Chien are shown in Figures 5.1 and 5.2. Although the results are represented in different ways and are computed from different

114

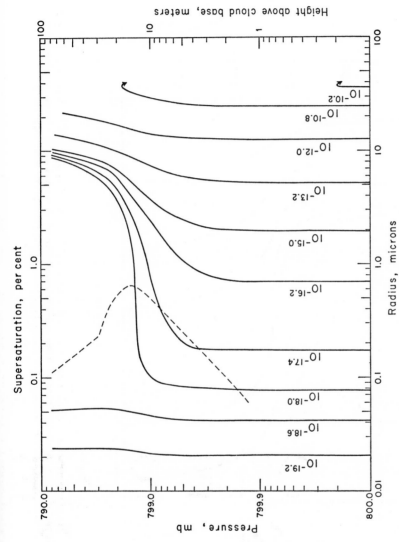

Fig. 5.1.—Growth of a population of droplets from a given distribution of NaCl nuclei as computed by Mordy (1959) for air ascending at 15 cm sec⁻¹. The trend of the supersaturation is given by the dashed curve in accordance with the scale at the top of the diagram. Each growth curve is labeled in terms of the number of moles of NaCl contained. The two larger ones develop fall speeds greater than 15 cm sec⁻¹ and so fall out, as indicated by the arrows. Note that the percentage of supersaturation, $(S - 1) \times 100$, increases at first, then decreases as vapor is used in condensation.

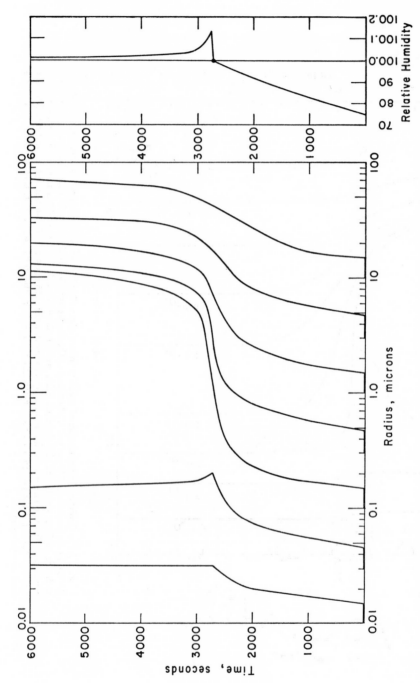

Fig. 5.2.—Growth of a population of droplets from a given distribution of NaCl nuclei as computed by Neiburger and Chien (1960) for a stratus cloud. The particles start with a dry size represented by the leftward extension of the foot of each curve. The relative humidity as represented in the graph on the right increases to 100 per cent in about 2700 seconds, then assumes supersaturated values.

NaCl nuclei distributions and cooling rates, they both show the same effect, namely a pronounced narrowing of the spectrum of sizes. The typical variations of the supersaturation were computed by the authors and inserted in the figures. The two sets of cumulative number concentrations as a function of size for the original nuclei and the final droplets are shown in Figure 5.3. The differences are mainly due to the broader nuclei spectrum chosen by Mordy as compared

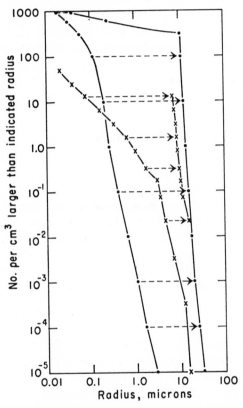

FIG. 5.3.—Initial and final distributions of particle and droplet sizes as computed by Mordy (points marked x) and by Neiburger and Chien (points with heavy dots). The dashed arrows connect initial with final sizes. The distributions are for the same data given in Figures 5.1 and 5.2.

with that used by Neiburger and Chien. The curves, when plotted in differential terms—number per cm³ per class interval of 1 cm radii—show a secondary peak for droplets of about 10μ radius. The primary maximum is in the smallest sizes. Mordy also finds this peak in terms of mass of condensed water contained in each size.

The two cases reproduced here represent slow rates of cooling, weak ascend-

117

ing motions, such as one might find in stratus clouds. When faster cooling, such as in cumulus clouds, is imposed the droplet size distributions are not appreciably different. The initial size spectrum of the nuclei has the greatest effect. When abundant giant nuclei are present, Mordy found that the liquid water has its greatest concentration on the sizes around 20μ. The secondary peak of the other distribution now becomes the primary maximum.

Computations of the kind just described pose the main problem of cloud physics. Since condensation produces a narrow spectrum of droplets of radii mainly of the order of 10μ, which are too small to fall, how does a cloud produce rain? Part of the answer is found in the giant nuclei which, while not numerous, could in some cases form enough drizzle droplets, which in turn could grow to raindrops by collision and coalescence with the main droplet population to produce appreciable rain. Other possible processes will be examined in a later section of this chapter.

Growth and Sublimation of Ice Crystals in Vapor

The growth of ice crystals and their sublimation to the vapor are treated in essentially the same way as growth and evaporation of liquid droplets. Since crystals are not spherical but rather in the form of hexagonal plates, stars, needles, columns, and related shapes, a radius cannot be assigned to them. The diffusion of vapor to the crystal is handled in a manner derived from the analogous situation in electricity—that of a current flowing to an object in an electric field spherically symmetric at infinity. The surface of the body is considered to be at a uniform potential V_0. The *capacity*, C, of the body enters into the problem. The current to the body is given by an application of Gauss's law as

$$i = \lambda \int_s \frac{\partial V}{\partial n} \, ds = 4\pi C\lambda (V_\infty - V_0), \qquad (5.27)$$

where λ is the electrical conductivity of the medium and ds is an element of surface area. The current is the flux of charge in the same sense that dM/dt is the flux of water vapor, and λ is analogous to D, the diffusivity of water vapor. The analogous expression in terms of vapor diffusion is

$$\frac{dM}{dt} = D \int_s \frac{\partial \rho_w}{\partial n} \, ds = 4\pi C D \lceil \rho_w - \rho_{w(s)} \rceil. \qquad (5.28)$$

This expression implies the existence of a density potential with a constant vapor density $\rho_{w(s)}$ at all points on the surface of the body. The capacity C is a geometrical factor which in the electrostatic units used here has the dimensions of length.[2] For a sphere, $C = r$, and for a circular disk, $C = 2r/\pi$.

---

[2] In the rationalized mks units C is in farads, and, for an isolated sphere, $C = 4\pi\epsilon r$ farads, where ϵ is the permittivity in farads m^{-1} of the medium, and r is in meters. For any medium $\epsilon = \kappa\epsilon_0$ where κ is the dielectric constant (~ 1 for air) and ϵ_0 is the permittivity in vacuo.

McDonald (1963) gave theoretical values of C for shapes most likely to be of interest in ice-crystal studies. In addition to a sphere and a circular disk, just given, he finds for a prolate spheroid of major and minor semi-axes a and b,

$$C = \frac{A}{\ln[(a+A)/b]},$$

where $A = (a^2 - b^2)^{1/2}$, and for an oblate spheroid

$$C = ae/\sin^{-1} e,$$

where $e = (1 - b^2/a^2)^{1/2}$ is the profile eccentricity. When a prolate spheroid becomes long and thin like a needle, $C = \ln(2a/b)$ as a limit, and for a very flat oblate spheroid C approaches $2a/\pi$, or the same as for a disk.

McDonald carried out tests with brass models of snow crystals suspended in the center of a walk-in Faraday cage, using a commercially available capacitance measuring assembly. It was found that hexagonal plates have about the same C as ideal thin disks of equal area. Increasing thickness did not affect the values by more than a few per cent. Needle-like cylinders had a capacitance between 70 and 80 per cent of that computed theoretically. Dendritic plates had values surprisingly close to those of hexagonal plates.

It is noted that equation (5.28) is the same as equation (5.4) except that C replaces r. In deriving the growth-sublimation equation for ice crystals it is not necessary to retrace the steps of equations (5.4) to (5.24) since C will everywhere replace r, L will become L_s for sublimation, and $p_e(T)$ for ice will replace $p_0(T)$ for the liquid. The saturation ratio $S = p/p_0(T)$ becomes $S_e = p/p_e(T)$. These alterations may be applied directly to (5.24), which then becomes, for an ice crystal,

$$\frac{dM}{dt} = \frac{4\pi C(S_e - 1)}{(L_s^2 m_w/KRT^2) + (RT/Dm_w p_e(T))} = \frac{4\pi C(S_e - 1)}{A+B}, \quad (5.29)$$

which is analogous to (5.24).[3]

Values of Terms in the Growth Equations

The two terms in the denominator of the growth equations are plotted as functions of temperature in Figure 5.4. In the work units used here and with C or r in cm, L is in ergs g^{-1}, R/m in ergs $°C^{-1} g^{-1}$, K in ergs $sec^{-1} cm^{-1} °C^{-1}$, D in $cm^2 sec^{-1}$, and p in dynes cm^{-2}, so that each term in the denominator has the dimensions cm sec g^{-1}. The terms a', b', A, B are of the same order of magnitude at temperatures near $0°$ C and sea-level pressures, but b' and B increase

---

[3] Note that L and L_s are expressed in work units, as before. Note also that T^2 in equation (5.29) is an approximation of $T_s T$, an approximation which is less reasonable here than in the condensation case since undercooling to $T \sim 250$ may be expected when ice crystals form. The released heat of sublimation may raise the ice-crystal temperature T_s to 273.

rapidly with decreasing temperature mainly because of the great sensitivity of the saturation vapor pressure to temperature. While a' and A contain terms that are functions only of the temperature, b' and B also vary with air pressure. The dependence on total pressure appears in the diffusivity D which bears the relation $D = D_0(p_0/p)$ thus making $B = B_0(p/p_0)$ and $b' = b_0'(p/p_0)$ at a given temperature. At 500 mb B has half the value it has at 1000 mb for the same temperature.

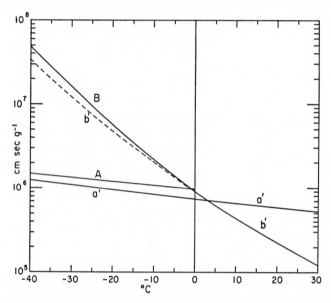

FIG. 5.4.—Terms in denominator of equations (5.24) (a',b') and (5.29) (A,B) as function of temperature at a pressure (applicable to b' and B) of 1000 mb. At other pressures, B and b' would have $p/1000$ths of the given values.

In Figure 5.4, b' and B are calculated for D at 1000 mb. The a' and b' which apply to condensation from the vapor to the liquid are extended into subfreezing temperatures in recognition of the importance of undercooling in clouds. The A is greater than the a' because $L_s > L$ while $B > b'$ because $p_0(T) > p_e(T)$ at temperatures below 0° C.

Values of K and D at various temperatures, the latter at a pressure of 1000 mb, are given in Table 5.1, taken from tabulated and computed data. (See, for example, *Smithsonian Meteorological Tables*, 1951, pp. 394–95.)

Langmuir, in a project report in 1944, formally published in 1961, called attention to a limitation of the diffusion theory of condensation when dimensions as small as the mean free path of the molecules are considered. In the lower atmosphere the mean free path of the water-vapor molecule is about 0.1μ. When droplets have radii in this range it is improper to assign the usual vapor pressure or vapor density at their surfaces. Langmuir introduced the expression equivalent to that used in developing (3.17) in chapter 3. In other words, kinetic theory is used to account for the escape and capture of molecules at the surface. Howell (1949) introduced Langmuir's modification as a correction to the diffusivity in the case of small droplets. Rooth (1957) carried the reasoning further and developed a very large correction factor, equal at $r = 10\mu$ to the correction that the Langmuir-Howell interpretation indicates is important only at $r < 0.5\mu$. It has been argued (Neiburger and Chien, 1960) that the values of the correction are too uncertain to justify modifying the diffusivity at normal cloud-droplet sizes, e.g., $r > 1\mu$. It is probable that similar arguments can be raised concerning the thermal conductivity.

TABLE 5.1

THERMAL CONDUCTIVITY K AND DIFFUSIVITY
OF WATER VAPOR IN AIR D AT
VARIOUS TEMPERATURES

Temperature (° C)	K erg cm^{-1} sec^{-1} ° C^{-1}	D (p = 1000 mb) cm^2 sec^{-1}
−40...........	2.12×10^3	0.169
−30...........	2.20	.183
−20...........	2.28	.197
−10...........	2.36	.211
0...........	2.43	.226
10...........	2.50	.241
20...........	2.57	.257
30...........	2.64	.273
40...........	2.70	0.289

In the scale in which wind measurements are made it is well known that the eddy diffusivity and the eddy conductivity replace D and K and that these eddy terms are orders of magnitude greater than the ordinary diffusivities and conductivities. If the droplets are small enough to be suspended, there is no air movement past them. Just as in the case of the ventilation effect, it is probable that one finds the eddy effect important only for large sizes. It is also probable that the experimentally determined ventilation factor includes eddy processes.

Marshall and Langleben (1954) have considered the possibility that the exchange of heat and moisture between the particle and its cloud environment is of a simple, direct convective nature such that the sensible heat transfer, instead of being given by (5.8) is of the form of (1.17), or

$$dQ = c_p(T_r - T) \tag{5.30}$$

in a unit mass of air, while the latent heat release is as in (1.41), or

$$dQ_L = L(w - w_r), \tag{5.31}$$

121

where w is the water-vapor mixing ratio. The heat balance, such as in equations (5.10) and (5.11), becomes

$$\frac{w - w_r}{T_r - T} = \frac{c_p}{L}. \qquad (5.32)$$

When this balance is substituted in the growth equation a slightly more rapid growth rate results, but the difference becomes important only near the temperatures and pressures giving the maximum growth. In this equilibrium K/D is replaced by $m_d c_p P/RT$, where m_d is the molecular weight of dry air and P is the total pressure; thus the air pressure and temperature variations are built in.

From available information and with consideration of the probable accuracies in various assumed processes, it appears that the basic growth equations, (5.22) and (5.29), with ventilation considered, are not due for a change until more reliable data are obtained.

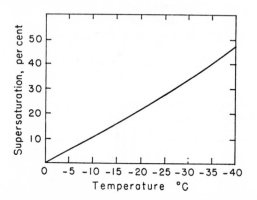

FIG. 5.5.—Supersaturation, $(S_e - 1) \times 100$, in per cent, over ice in an undercooled liquid cloud at water saturation.

Growth of Ice in Water Clouds

Since ice crystals frequently grow in undercooled liquid clouds, one is likely to find the vapor pressure in equilibrium with undercooled water. The saturation ratio with respect to ice would then be $S_e = p_0/p_e$, which is a function of temperature only. The quantity $S_e - 1$ increases almost linearly from 0 at $0°$ C to 0.475 at $-40°$ C, as shown in Figure 5.5.

In Figure 5.6 are plotted growth-rate curves for a snow crystal having $C = 1/4\pi$ in a cloud in liquid-vapor equilibrium or, in other words, growth rates in terms of $(S_e - 1)/(A + B)$ for $S_e = p_0/p_e$. To obtain the true growth for any other C, the values need only be multiplied by $4\pi C$. It is noted that a maximum occurs at $-14.25°$ C at 1000 mb and at $-16.75°$ C at 500 mb. The maxima come about because the denominator in equation (5.29) increases proportionately less rapidly with decreasing temperature than does p_0/p_e down to these temperatures, then more rapidly at the lower temperatures. The pressure effect on B causes the two curves in Figure 5.6 to be quite distinct, and it is apparent that the growth rate is strongly affected. This effect is brought about, as has been seen, by the increase in the diffusivity of water vapor with decreasing pressure.

A realistic way of looking at the temperature and pressure effects in natural clouds is to consider the lapse rate or P,T distribution. In active clouds the lapse rate usually is close to the saturation-adiabatic equilibrium. In Figure 5.7 four saturation-adiabatic distributions identified by equivalent-potential temperature are considered, ranging from $\theta_E = 350$ for typical U.S. summer conditions to $\theta_E = 290$, approximating winter snow clouds. The rates of growth, given

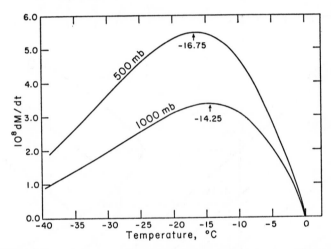

Fig. 5.6.—Growth rates as a function of temperature of an ice crystal with $C = 1/4\pi$ in a water-saturated cloud at two pressures. The temperature for maximum rate of growth is indicated on each curve.

in terms of $(1/4\pi C)(dM/dt)$, are plotted in four accompanying curves corresponding to the four P,T distributions, and with $S_e = p_0(T)/p_e(T)$. It is seen that the growth is fastest in the warmest atmosphere. This situation results from the fact that the subfreezing temperatures occur higher up in the atmosphere—at lower pressures—so the pressure effect on B, through D^{-1}, is greater. The point at which the maximum growth rate occurs is slightly colder the warmer the atmosphere, as shown by the dashed lines which connect the growth-rate curves to the P,T curves thence to the abscissa.[4]

[4] It is interesting to note that a' and b' are always smaller, respectively, than A and B, so for the same supersaturations, the condensation growth rates would be greater than those shown here. But the point is that such large supersaturations do not occur with respect to water, the dM/dt for ice crystals being greater because of the larger

123

Growth curves in terms of size can be constructed from an integration of the fundamental equation if the crystal shape is known so that M and C are expressed in terms of r. For example, a common form is the hexagonal plate for which the capacity can be taken with reasonable approximation as that of a disk, $C = 2r/\pi$. The dimension r for the hexagon can be considered as the dis-

---

numerator. Since condensation predominates over ice-forming processes at subfreezing temperatures warmer than about $-20°$ C, it appears that the rate of vapor flux to the cloud particle is not the determining factor for ice formation versus condensation.

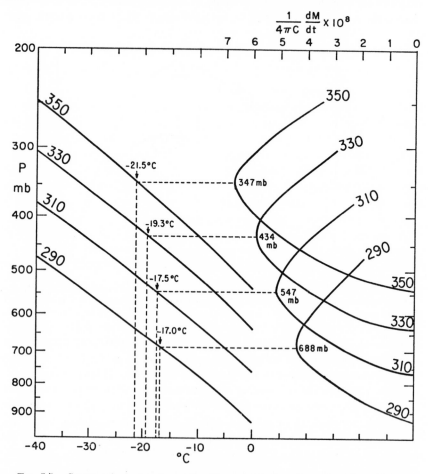

Fig. 5.7.—Comparative rates of growth of ice crystals in water-saturated environments having P,T distributions corresponding to four different saturation-adiabatic lapse rates, identified by equivalent-potential temperatures. Rates are given by scale at top of figure in terms of $(1/4\pi C)(dM/dt)$. Maxima are marked in terms of temperature and pressure by dashed lines.

tance from the center to each of the six corners, giving the area as $s = 2.6\,r^2$. Let us make a computation assuming that the thickness is $z = 2r/3$ (Reynolds, 1952) and the density $\rho_e = 0.5$. Then

$$M = 0.5 \times 2.6r^2 \times 2r/3 = \rho_e sz$$

and

$$\frac{dM}{dt} = \rho_e \left(s\frac{dz}{dt} + z\frac{ds}{dt} \right) = 7.8r^2\frac{dr}{dt} = \frac{4\pi C(S_e - 1)}{A + B}; \quad (5.33)$$

and, with C substituted for a disk,

$$r\frac{dr}{dt} = 3.077\frac{S_e - 1}{A + B} = \frac{3.077}{4\pi C}\frac{dM}{dt}. \quad (5.34)$$

Integrated, as in (5.23), this becomes

$$r^2 = r_0^2 + 6.154\frac{S_e - 1}{A + B}(t - t_0) = \frac{6.154}{4\pi C}\frac{dM}{dt}\delta t. \quad (5.35)$$

Curves of this type are shown in Figure 5.8, based on the maximum values of $(1/4\pi C)(dM/dt)$ given in Figure 5.7 for $\theta_E = 350$ and $\theta_E = 290$ and an initial r_0 of 1μ. It is seen that the growth is very rapid at first but fairly slow after a few minutes. For comparison, two curves of growth by collision-coalescence are included, for drops of initial radius 30μ and 50μ. The comparison illustrates the fact that diffusion growth is important only as an initiating mechanism and that once a size is attained representing an appreciable terminal velocity, coalescence becomes the predominating mechanism.

Reynolds (1952) measured the rates of growth of crystals in an undercooled cloud in a large laboratory cold box after dry-ice "seeding." He found the growth to proceed in a manner closely approximating computations made by Houghton (1950) from the usual equations.

Marshall and Langleben (1954) have considered the case of a crystal in an undercooled cloud in which the vapor tension of the liquid droplets is not found at infinity, as assumed in the development of the growth equations (5.29), etc. Droplets in a spherical shell of finite radius are considered as enclosing the ice particle. The expression for evaporation of each droplet, with heat exchange neglected, is given as in (5.3) in the form

$$-dM_d/dt = 4\pi r_d D[\rho_{w(d)} - \rho_w], \quad (5.36)$$

where $\rho_{w(d)}$ is the vapor density corresponding to the vapor tension of the droplet and ρ_w is the vapor density at a suitable distance away from it. For a unit volume of the thin shell of droplets, the evaporation is the total for all the droplets or

$$4\pi \sum_1 r_d D[\rho_{w(d)} - \rho_w], \quad (5.37)$$

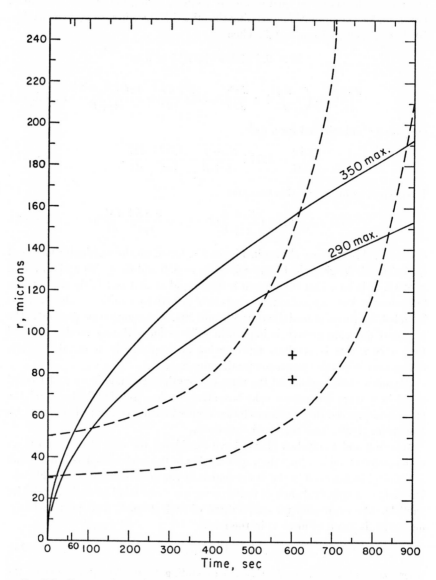

FIG. 5.8.—Integrated growth curves of a hexagonal-plate crystal having a radius (center to corners) of 1μ initially, growing at the maximum rates given in Figure 5.7 for $\theta_E = 350$ and $\theta_E = 290$, compared with growth (dashed curves) of drizzle drops. The drops are considered at $t = 0$ to be in a cloud containing 1 g m^{-3} of liquid water at $-3°$ C reaching 470 mb after 8 minutes ($t = 480$) where the temperature is $-10°$ C and the liquid-water content is 2 g m^{-3}. The equivalent melted radii of the crystals at 10 min are given by the two crosses.

where the summation is in a unit volume. The contribution of vapor from such a shell of radius x_1—distance x_1 from the crystal—and of thickness Δx would be

$$\Delta \frac{dM}{at} = 4\pi x_1^2 \Delta x \left\{ 4\pi \sum_1 r_d D [\,\rho_{w(d)} - \rho_w\,] \right\} \tag{5.38}$$

or, considering any distance x from the ice particle,

$$\frac{d}{dx}\left(\frac{dM}{dt}\right) = 4\pi x^2 \left\{ 4\pi \sum_1 r_d D [\,\rho_{w(d)} - \rho_w\,] \right\}. \tag{5.39}$$

But the flow of mass by diffusion of vapor along the spherically symmetric gradient of ρ_w—across the spherical vapor-density surfaces—is

$$\frac{dM}{dt} = -4\pi x^2 D \frac{d\rho_w}{dx}; \tag{5.40}$$

so the left-hand side of (5.39) becomes

$$\frac{d}{dx}\left(-4\pi x^2 D \frac{d\rho_w}{dx}\right) = -4\pi D \left(x^2 \frac{d^2\rho_w}{dx^2} + 2x \frac{d\rho_w}{dx}\right), \tag{5.41}$$

and the equality (5.39) becomes, after dividing both sides by $4\pi x^2 D$,

$$\frac{d^2\rho_w}{dx^2} + \frac{2}{x}\frac{d\rho_w}{dx} = 4\pi \sum_1 r_d [\,\rho_w - \rho_{w(d)}\,] \tag{5.42}$$

or

$$\frac{d^2\rho_w}{dx^2} + \frac{2}{x}\frac{d\rho_w}{dx} - [\,\rho_w - \rho_{w(d)}\,]\left(4\pi \sum_1 r_d\right) = 0. \tag{5.43}$$

This equation is solved by substituting a non-dimensional scale factor in terms of length

$$\lambda = \left(4\pi \sum_1 r_d\right)^{1/2} x = kx. \tag{5.44}$$

It is seen that $k^2 = 4\pi \Sigma_1 r_d$ and that k has the dimensions of inverse length. When k is zero, x is infinite, and we have the classical growth equation

$$\frac{dM/dt}{4\pi D [\,\rho_{w(d)} - \rho_{w(e)}\,]} = C \tag{5.45}$$

($= r_e$ for an ice sphere). By substituting and dividing through by k^2, the equation becomes

$$\frac{d^2\rho_w}{d\lambda^2} + \frac{2}{\lambda}\frac{d\rho_w}{d\lambda} - [\,\rho_w - \rho_{w(d)}\,] = 0. \tag{5.46}$$

The boundary conditions applied to solve this differential equation are $\rho_w = \rho_{w(d)}$ at $\lambda = \infty$ and $\rho_w = \rho_{w(e)}$ at $\lambda = \lambda_e$, where the subscript e refers to the crystal surface, actually assumed to be spherical to make a solution possible. The solution of (5.46) is given as

$$\rho_w = \rho_{w(d)} - [\rho_{w(d)} - \rho_{w(e)}]\frac{\exp(-\lambda)}{\lambda}\frac{\lambda_e}{\exp(-\lambda_e)}, \qquad (5.47)$$

which may be simplified by substituting $\lambda = \Lambda\lambda_e$ (or $x = \Lambda r_e$), resulting in

$$\rho_w = \rho_{w(d)} - [\rho_{w(d)} - \rho_{w(e)}]\Lambda^{-1} \exp[-\lambda_e(\Lambda - 1)]. \qquad (5.48)$$

The values of ρ_w obtained from this expression can be substituted into the diffusion equations such as (5.40). Without going into details we may note that Marshall and Langleben find that at the surface of a spherical crystal we have, in place of the classical equation (5.45), the rate of growth given by

$$\frac{dM/dt}{4\pi D[\rho_{w(d)} - \rho_{w(e)}]} = r_e(1 + kr_e). \qquad (5.49)$$

In other words, a quantity kr_e^2 is added to the radius factor in the growth rate. For a non-spherical shape, an r_e or similar dimension as a function of electrical capacity may be introduced.

What values can k have? In an undercooled stratus cloud the liquid water content may be on the order of 0.3 g m^{-3}. If the droplets have a mean radius of 2μ, k would be of the order of 4 or 5 cm^{-1}. In heavy cumulus, cumulonimbus or nimbostratus k might have a value around 10 cm^{-1}. By taking the ratio of (5.49) to (5.45), one can compare the rates of growth under the assumed cloud conditions with the classical case. One finds

$$\frac{r_e(1 + kr_e)}{r_e} = 1 + kr_e. \qquad (5.50)$$

If r_e is 0.05 cm and k is 5 cm^{-1} the growth rate is 1.25 times the classical rate. In the beginning states, where r_e may be of the order of microns, the correction is almost negligible. Since dr/dM is inversely proportional to the square of the mass, therefore proportional to r^{-6}, an increase in mass of large particles means very little increase in their size, so the Marshall and Langleben correction is less important in terms of size growth.

The Bergeron Process

The growth of ice at water saturation, therefore ice supersaturation, as described in the previous section, was recognized as important in natural clouds by the Swedish meteorologist Tor Bergeron in 1928 and was outlined by him (1934) in qualitative detail at an international geophysical congress in Lisbon in 1933. Bergeron concluded that, except for some tropical showers and the usual drizzle situations, precipitation was *initiated* only by the process of ice-crystal growth in undercooled clouds. In recent years the relative importance of ice initiation versus an all-liquid process has been fairly well documented from radar data and meteorological research flights, and the exceptions to the Bergeron process have been found to be more common than was at first realized.

It has been determined (Byers and Hall, 1955) that the ice phase is never required for initiating precipitation over tropical oceans, and probably is not a requirement over humid tropical land masses either. In convective clouds with bases low enough so that a cloud depth of nearly three kilometers exists between the base and the melting level, precipitation-size drops have an opportunity to develop within the liquid portion. This is a common situation in large summer cumuli developing into cumulonimbi in the eastern half of the United States (Battan, 1953; Braham *et al.*, 1957). The University of Chicago flights in the Central States have shown that on certain days small, warm convective clouds initiate precipitation, in many ways similar to the shower-producing cumuli of the tropical ocean areas. In general, though, the continental cumuli grow much bigger and taller before producing a radar echo.

In the 1960–63 summer research flights in Missouri, Braham (1961, 1964) and Koenig (1963) found that solid pellets developed in the summer cumuli as they penetrated through the −5 to −10° C layer. The density and shape of the ice particles and the cloud history suggested that they were formed by the freezing and shattering of precipitation-sized drops that had been grown by liquid condensation plus coalescence as they were carried in the updraft.

In convective clouds in the arid Southwest where bases were only a kilometer or so below the 0° C isotherm, radar studies indicated that ice processes could very well account for the initial radar echo (Braham *et al.*, 1951; Reynolds and Braham, 1952; Morris, 1957; Braham, 1958; Ackerman, 1960). From the number of cases of pellets rather than snowflakes observed in University of Chicago flights through the initiating levels of these clouds, it might be inferred that even there the ice in a number of cases formed from the freezing of drops of precipitation size. In radar studies in Arizona many of the first echoes to appear in building clouds were found straddling the 0° C level.

In the large-scale cyclonic storms of middle and high latitudes, especially in the colder seasons, where ascents are gradual and in layers, the Bergeron process unquestionably operates. These are the kinds of rainstorms familiar to Bergeron in Northern Europe, and they occur under the same circumstances in North America. In 60 research flights in general rainstorms of western Washington, Hall (1957) observed the Bergeron process operating with a variety of distributions of ice crystals and undercooled droplets in the cloud systems. He noted frequent cases where there was an upper and a lower cloud deck, "the lower containing the primary water supply and the upper furnishing ice crystals which fell into the lower deck and rapidly increased in size." In Figure 5.9 two cases of non-layered nimbostratus typical of those found by Hall are illustrated. In *A* the more-or-less classical situation is seen. In *B* liquid cumuli are embedded in the bottom of a snow cloud, their updrafts apparently sufficient to prevent appreciable entrainment of surrounding crystals. Liquid droplets also are indicated in the main cloud top. It is probable that this part of the cloud, being quite diffuse, is slightly subsaturated with respect to water, and so the ice supersaturation is insufficient to produce crystals.

It should be emphasized that the Bergeron process is concerned with the *initiation* of precipitation. The water that reaches the ground is acquired predominantly by collision and coalescence—clumping and riming in the case of snow crystals—as the drops or crystals move relative to the smaller cloud particles, often picking up most of the water in the low part of the cloud where

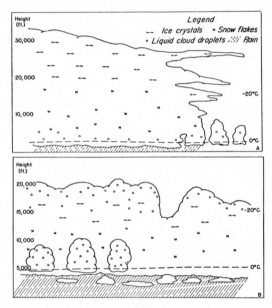

FIG. 5.9.—Schematic distributions of ice and liquid forms in clouds of winter storms in western Washington as given by F. Hall (1957).

the liquid-water content may be high.[5] Cunningham (1951) reported on detailed flight studies of general storm clouds in New England. Later (Cunningham and Atlas, 1954) aircraft measurements were combined with quantitative analyses of radar reflectivities to calculate growth rates of hydrometeors occurring in various parts of general cyclonic snow-rain systems. From the detailed study of five big storms the authors conclude: "Near the center of active storms three-fourths of the moisture came from the low-level clouds, largely underneath the frontal surface." The mass growth rates were found to increase more or less logarithmically from the generating level to the lower clouds, then

[5] In the cores of cumulonimbus updrafts the water content may be higher in the upper part of the cloud as prescribed by the parcel-ascent theory.

peaked sharply within them to values 100 to 1000 times the initial rate. These accretions of water could have occurred only by coalescence.

All evidence in general cyclonic storms (see also Byers, 1959; McClain and Omoto, 1962) points to the importance of mesoscale convective systems contained in the general cloud assemblage, and of the liquid water in the lower cloud layers. For ample rains to occur in extratropical cyclones, low-level convergence and ascent processes must continually supply the lower convective or layer clouds. Frontal upslope effects will produce only light precipitation if there are no wet clouds in the lower, polar air.

Growth Habit of Ice Crystals

Persons living in snowy climates are aware of the myriad delicate forms ice crystals can take, both in their formation on surfaces and in falling snow. Only the latter need be considered under the heading of cloud physics, but the forms observed in frost deposits, glaciers, ponds, and streams furnish clues concerning crystal structure in general. In ice and frost structures on ground objects the orientations and forces of molecular scale on and around the surface object often play a more important part in determining the growth habit than does the meteorological environment. In clouds and precipitation the ambient air conditions are the dominant factors.

The late Japanese physicist Ukichiro Nakaya devoted thirty years of painstaking work to the study of snow and ice crystals, including the taking of thousands of photomicrographs. His book *Snow Crystals* (1954), delayed in publication ten years by World War II, is a classic.[6] His work has been augmented by other studies in laboratories in various parts of the world, and there now exists a considerable body of literature on the subject.

The solid hydrometeors observed in the atmosphere can be classified relatively simply. The pristine crystal forms grow in a vapor diffusive field without direct contact with the other hydrometeors. More complex shapes develop from attachment of other crystals, from clumping, riming, fracturing, and other events in their life history. The pristine crystals are growths upon the basic hexagonal, or occasionally trigonal, pyramids and bipyramids. There are three fundamental structures: needles, columns, and plates, the first two representing development mainly in the principal axis and the last predominantly in the lateral axis. The needles occur singly or in bundles. Among columnar forms are included pyramids and "bullets." The plates range from simple hexagons, through sectored hexagons and stars to the highly branched dendrites. Apparently under exposure to successively different diffusive fields, combinations such as capped columns (columns with plates at the ends), spatial (three-dimensional) dendrites, and

---

[6] The art of snow microphotography was beautifully displayed in a book by Bentley and Humphreys published in 1931 (see references).

other combined forms are found. Microphotographs of some elementary forms are seen in Plate 5.1.

Clumping and riming produce such forms as large, cottony snowflakes, thick plates, rimed crystals, low-density snow grains, dense pellets, and miscellaneous irregular particles. Some of the pellets, hailstones, etc., result from the freezing of drops. The World Meteorological Organization (1956) has published definitions of hydrometeors, and the International Association of Scientific Hydrology (1954) has attempted a classification of all ice forms observed in nature.

Nakaya produced snow crystals on rabbit hairs in a temperature-controlled chamber at various supersaturations. He developed a diagram, represented in Figure 5.10, showing how the different growth habits depend on temperature and, to some extent, on the supersaturation. The boundaries separating the different forms are entered empirically. This diagram has been verified generally also in studies of snow forms from natural clouds in which the temperatures were known. In his figure, Nakaya superimposed a curve representing the supersaturation with respect to ice that would prevail in an undercooled cloud at water saturation. The subject can be further elucidated by considering the ice-crystal growth function in such undercooled clouds, and this has been done in Figure 5.10 by plotting it for 1000 mb under Nakaya's diagram.

If the diffusive growth rates represented by the curve in Figure 5.10 are compared with the snow-crystal forms, it is seen that the most intense diffusion is associated with the range of the dendrites, with plates predominating on the shoulders of the curve. (The plates at low supersaturations around $-15°$ C need not be considered in the discussion of this highly supersaturated case of a water-saturation environment.) The scroll or cup is found in a narrow region on the high-temperature side, then at about -2 to $-8°$ C the needles represent the least intense diffusion. Marshall and Langleben (1954) noted that if lines of constant growth factor in a water-saturation environment are substituted for Nakaya's boundaries, the data are accommodated about equally as well. Gold and Power (1954) found evidence that the ranges of the different forms drifted toward lower temperatures with height in accordance with the trend of the maximum of the growth curve toward lower temperature with decreasing pressure (cf. Fig. 5.6).

Pyramidal prisms are of special interest because they form characteristically at the low temperatures ($< -20°$ C) found in cirrus and cirrostratus clouds. *Virga* from these clouds, consisting of falling streamers of ice crystals, are commonly seen, suggesting that the prisms, either singly or in clusters, have a higher settling speed (terminal velocity) than other forms because of either shape, density, size, or combinations of these. This phenomenon, occurring as it does over lower undercooled clouds of cyclonic systems, provides a means of starting the Bergeron process. A pyramid photographed by Kumai (1961) is shown in Plate 5.2.

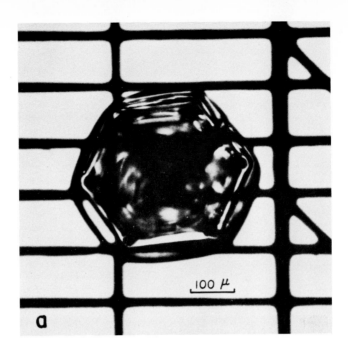

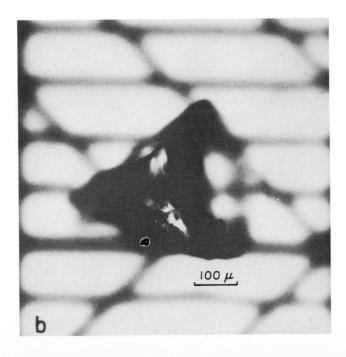

PLATE 5.1.—Microphotograph by M. Kumai (1961) of pyramid crystal collected in natural snowfall: *a*, base plane; *b*, side view.

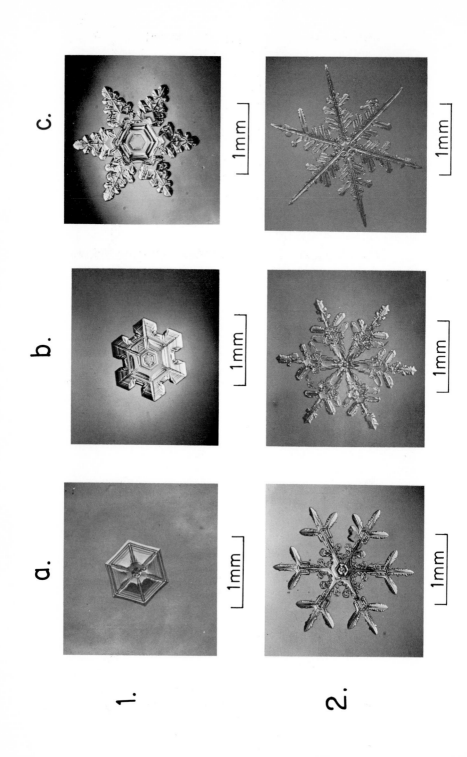

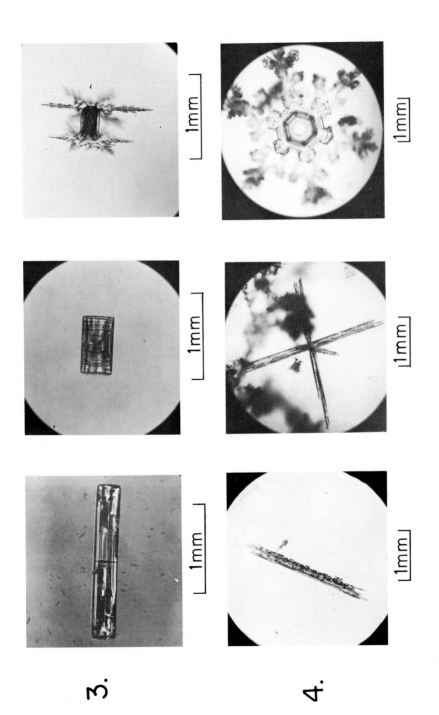

3.

4.

PLATE 5.2.—Microphotographs of some of the elementary forms of ice crystals: 1 *a*, *b*, *c*, hexagonal plates showing growth at corners in *b* and *c*; 2 *a*, *b*, *c*, dendritic forms; 3 *a*, *b*, columns; 3 *c*, capped column, shown in end view in 4 *c*; 4 *a*, *b*, needles. (Photographs 1 *a* through 3 *a* by Mr. Kazuhiko Itagaki; 3 *b* through 4 *c* by the late Dr. Ukichiro Nakaya.)

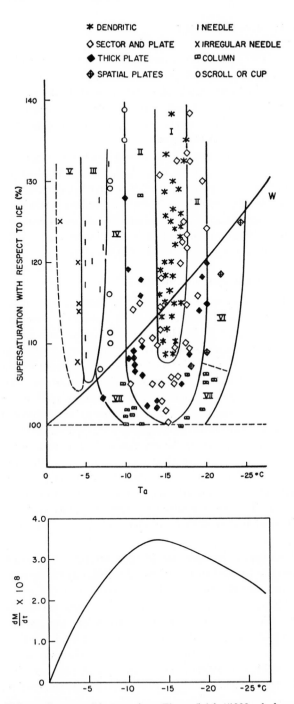

Fig. 5.10.—Nakaya diagram with curve from Figure 5.6 for 1000 mb drawn to the same temperature scale. The curve in the upper diagram marked W gives the supersaturation with respect to ice in a water-saturated environment.

Weickmann (1945) found from photomicrographs of crystals collected in a number of flights in cirrus and cirrostratus clouds that prisms predominated. At temperatures in these clouds of $-20°$ to $-30°$ C the crystals were solid, single, or twin prisms while at temperatures colder than $-30°$ C partly hollow prisms, often aggregated in clusters, were common. The clusters were joined at the pyramidal apexes or at the faces.

Interesting growth effects are seen in the laboratory when an electric field greater than about 500 volt cm^{-1} is applied. Bartlett *et al.* (1963) found that rapid growth of long, unbranched needles occurred under these circumstances. The needles grow 10 to 100 times faster than ordinary needles. They can be made to grow on other forms, such as dendrites, at temperatures characteristic of the basic crystal rather than the needle. It is interesting to note that electric fields of the required magnitude exist in thunderclouds.

Replicas and Laboratory Experiments

A standard way of collecting and studying specimens of snow crystals is to make casts or replicas of them. The most commonly used method is that developed by Schaefer (1946) in which a plastic having the trade name Formvar —a polyvinal acetal resin—is dissolved in ethylene dichloride to provide the replicating material. The solution is kept at a temperature of about $-5°$ C while a microscope slide is immersed in it for about 30 sec. The crystals are allowed to settle on the slide where they become submerged in the film of solution. The solvent evaporates in a few minutes at the subfreezing collection temperature leaving the plastic to harden as a transparent cast. When the slide is exposed to room temperature the crystals melt, the water diffusing through the plastic and evaporating. Detailed replicas remain in the plastic. Crystals may be selectively collected by letting them fall on black cloth, choosing those of interest with the aid of a hand lens, and picking them up by touching them with the coated slide. Most modern photomicrographs are made of replicas rather than the original crystals for obvious reasons of convenience.[7]

The ordinary food freezing box is standard equipment for studying ice crystals. A cold diffusion chamber providing a strong temperature and water-vapor gradient can be used to study growth habit. Crystals grown on filaments can be raised or lowered through various ranges of temperature and the change in crystal habit noted. As the temperature rises from, let us say, $-20°$ C at low supersaturation to $-15°$ C at water saturation, one would observe first columns, then plates capping the columns followed by dendritic growth of the plates. Finally, as the temperature rises farther, needles would begin to branch off. Hallett and Mason (1958) have reported, with photographic documentation, several such sequences. Isono (1957) has demonstrated different rates of

[7] For a more complete discussion of replica methods see the reference to Kobayashi (1955).

growth on edges and faces of crystals under varying conditions of temperature and humidity.

The nucleus of an ice crystal can be studied by collecting the crystal on a collodion-coated slide and allowing it to sublimate, while noting by means of a background grid the location of the geometric center, which is the last part of the crystal to sublimate. When this spot is examined later in the electron microscope the nucleus is found there. This method has been highly developed by Kumai (1961).

By means of the analogy between electric current and vapor flux, the growth in various crystal axes can be simulated in an electrolytic tank. A model of a crystal made of conducting material is immersed in the center of a tank of conducting liquid such as a weak acid. A potential is applied to the walls of the tank, and the flow of current to the model is measured. If the relative dimensions are appropriate one can assume equipotential surfaces at spherical infinity from the model. Marshall and Langleben (1954) constructed a double or triple "sandwich" model of tabular forms, separated by insulators but with one disk exposed. The current was measured individually to each model and the flow to edges and points compared with that to the exposed face, etc.

One might expect that the stronger the applied voltage the greater would be the tendency for the current to flow to the points, and that at lower potentials the flow would be more toward the faces. The effects are not quite so simple. One difficulty is that the analogy is not between constant charge density and constant vapor density at the surface of the object but between voltage and vapor density. Marshall and Langleben suspected that in real crystals a kind of Kelvin curvature effect produces high vapor densities at the points and edges.

The problem is one of determining how the balance

$$\frac{A+B}{4\pi(S_e-1)}\frac{dM}{dt}=C \tag{5.51}$$

varies at constant pressure as S_e and the temperature change. Since C is a function of shape, one assumes that the Nakaya boundaries represent critical values of S_e and T at any given atmospheric pressure. Considering the balance in the reverse sense, that is, the effect on the other terms of a change in the capacity C, one finds that in going from a hexagonal plate to a dendrite a reduction in C occurs. To change the form while preserving the balance expressed by (5.51), one must change the temperature, so the terms on the left which are functions of temperature will have different values. The measurements by McDonald of C on various models in a Faraday cage have already been mentioned.

Artificial Nucleation

The first experiments in changing an undercooled cloud to ice crystals in the laboratory and in the atmosphere (Schaefer, 1946) were accomplished through lowering the temperature by dropping pellets of solid CO_2 into the cloud. Each pellet momentarily cooled a path of droplets to a temperature where freezing would occur homogeneously or otherwise, as the purity of the droplets required. In this way tremendous numbers of ice crystals could be formed with a small quantity of "dry ice," and these in turn would infect the cloud by spreading in turbulent mixing.

Shortly after Schaefer's experiments with dry ice, Vonnegut (1947) discovered the usefulness of the epitaxially favorable crystals of silver iodide. They are produced as a smoke from generating systems that have undergone improvements over the years. A common method is to burn acetone containing silver iodide in solution. To prepare the solution, sodium or potassium iodide is used with silver nitrate ($AgNO_3 + KI \rightarrow AgI + KNO_3$).

While dry ice can produce ice crystals in an undercooled cloud at any subfreezing temperature up to approximately $0°$ C, silver iodide has a nucleating "threshold" of $-4°$ C or lower. In practice, the AgI has the advantage that it can be dispersed in smoke particles fine enough to become aerosols that diffuse with the air currents. Most commercial cloud seeding is attempted from generators located on the ground, whence the particles are supposed to be carried by the wind, turbulent mixing, and updrafts to the undercooled portions of the cloud.

Other substances that have been reported to serve as artificial nuclei about as well as AgI are PbI_2, CuS (cupric sulfide), CuO, Cu_2O, BiI_3, and some others (Fukuta, 1958), but AgI has received the most attention. Experiments with some organic materials have also been reported (Braham, 1963; Head, 1961, 1962*a, b;* Komabayasi and Ikebe, 1961; Krasikov, 1961; Langer and Rosinski, 1962; Power and Power, 1962).

Katz (1961, 1962) performed cloud-chamber experiments to determine the ice-nucleating activities of various aerosol substances. Silver iodide was found to be the most active at nearly all temperatures, but cupric sulfide (CuS) followed closely. His experiments pointed out, however, that the "threshold temperature" has only limited meaning, since the nucleating activities of all the aerosols, especially AgI and CuS were strongly dependent on temperature, decreasing by an order of magnitude or more with each degree of increase in temperature at the warmer end of the range. There was also an appreciable effect of humidity, especially at the higher temperatures; for example, the AgI activity at $-7°$ C was about 10 times greater at water saturation than at ice saturation, at $-15°$ C about three times greater.

If a particle acts as an ice-nucleating agent in a humid atmosphere it probably does so by first adsorbing a film of liquid water. A number of laboratory experiments on the water adsorption of silver iodide have been tried. However, Tompkins *et al.* (1963) found that the potassium iodide normally used in the preparation of silver iodide results in the formation of a double salt, probably AgI·3KI, in the presence of water vapor at a fairly low saturation ratio.

Like other silver halides, silver iodide is photochemically sensitive, and it is found that its ice-nucleating ability is diminished by exposure to ultraviolet light. The effect was first reported by Reynolds *et al.* (1951) and confirmed in various aspects by Inn (1951), Vonnegut and Neubauer (1951). Smith *et al.* (1955, 1956, 1958) found that the reduction in nucleating activity varied from 10 to 10^6 per hour in sunlight, depending on the type of generator and method of preparation of the material. Fletcher (1959), from a simple theoretical model, assessed the photolysis process as depending primarily on the size distribution of the smoke particles and their chemical composition. The smaller smoke particles are most readily affected.

Seeding of undercooled liquid clouds to produce ice crystals is aimed at starting the Bergeron process. In this way it is possible that precipitation can be initiated. The action is clearly demonstrated in the case of seeding of thin, undercooled stratus cloud layers by dry ice or silver iodide put into the clouds from an airplane. The ice crystals, reflecting sunlight in a manner different from liquid droplets, show up quite clearly in a swath along or below the flight path of the airplane. In a later stage enough snow may fall from the cloud so that the swath becomes cleared of cloud. Turbulent diffusion of ice particles may result in a cleared path from a single airplane of more than a kilometer in width.

In the chaotic motions and short lifetimes of convective clouds the effects of seeding are difficult to demonstrate. In the changing aspect of each cloud one is never sure as to whether glaciation was caused artificially or naturally. It is equally difficult to assign artificially induced or entirely natural causes to any precipitation which might fall from such clouds. Furthermore, many cumulus clouds seem capable of producing small raindrop sizes without the aid of the Bergeron process, as pointed out earlier in this chapter and as will be detailed further in chapter 6.

Proof of increases of precipitation over a period of time or in specific instances reduces to a quite complicated statistical problem. Randomized statistical approaches are necessary in order to determine the probability that the observed conditions could have occurred naturally. The great natural variability of precipitation in most locations makes the discovery of unnatural effects extremely difficult. In many parts of the world, including the United States, public policy concerning "weather control" is often guided by claims of cloud-seeding success based on evidence so questionable as to seem farcical to a sophisticated statistician. The statistical design and evaluation of such experiments will not be covered here.

REFERENCES

ACKERMAN, B. (1960), in *Physics of Precipitation*, ed. H. WEICKMANN (Amer. Geophys. U. Monogr. No. 5), p. 79.

BARTLETT, J. T., VAN DEN HEUVEL, A. P., and MASON, B. J. (1963), *Zeitschr. angew. Math. Phys.*, **14**, 599.

BATTAN, L. J. (1953), *J. Met.*, **10**, 311.

BENTLEY, W. A. and HUMPHREYS, W. J. (1931), *Snow Crystals* (New York: McGraw-Hill Book Co.).

BERGERON, T. (1928), *Norske Vid. Akad., Geofys. Pub.*, **5**, No. 6.

—— (1934), *Mem. Internat. Assn. Met.*, Lisbon, 1933.

BRAHAM, R. R., JR. (1958), *J. Met.*, **15**, 75.

—— (1961), *Internat. Assn. Met. and Atmos. Phys., Proc. Conf. Cloud Phys., Canberra and Sidney, 1961*.

—— (1963), *J. Atmos. Sci.*, **20**, 563.

—— (1964), *ibid.*, **21**, 640.

BRAHAM, R. R., JR., BATTAN, L. J., and BYERS, H. R. (1957), *Amer. Met. Soc. Met. Monogr.*, **2**, No. 11, p. 67.

BRAHAM, R. R., JR., REYNOLDS, S. E., and HARRELL, J. H., JR. (1951), *J. Met.*, **8**, 416.

BYERS, H. R. (1959), in *The Atmosphere and the Sea in Motion* (New York: Rockefeller Institute Press), p. 400.

BYERS, H. R., and HALL, R. K. (1955), *J. Met.*, **12**, 176.

CUNNINGHAM, R. M. (1951), *Bull. Amer. Met. Soc.*, **32**, 334.

CUNNINGHAM, R. M., and ATLAS, D. (1954), *Roy. Met. Soc., Proc. Toronto Met. Conf. 1953*, p. 276.

FLETCHER, N. H. (1959), *J. Met.*, **16**, 249.

FRÖSSLING, N. (1938), *Beitr. z. Geophys.*, **52**, 170.

FUKUTA, N. (1958), *J. Met.*, **15**, 17.

GOLD, L. W., and POWER, B. A. (1954), *J. Met.*, **11**, 35.

GUNN, R., and KINZER, G. D. (1949), *J. Met.*, **6**, 243.

HALL, F. (1957), *Amer. Met. Soc. Met. Monogr.*, **2**, No. 11, p. 33.

HALLETT, J., and MASON, B. J. (1958), *Proc. Roy. Soc. London, Ser. A*, **247**, 440.

HEAD, R. B. (1961), *Nature*, **191**, 1058.

—— (1962a), *J. Phys. Chem. Solids*, **23**, 1371.

—— (1962b), *Nature*, **196**, 763.

HOUGHTON, H. G. (1950), *J. Met.*, **7**, 363.

HOWELL, W. E. (1949), *J. Met.*, **6**, 134.

INN, E. C. Y. (1951), *Bull. Amer. Met. Soc.*, **32**, 132.

INTERNAT. ASSN. SCI. HYDROL., Commission on Snow and Ice, Associate Committee on Soil and Snow Mechanics, Snow and Ice Research (1954), Tech. Memorandum No. 31, Ottawa (Nat'l. Res. Council Canada).

ISONO, K. (1957), *Met. Soc. Japan 75th Anniv. Vol.*, p. 31.

KATZ, U. (1961), *Zeitschr. angew. Math. Phys.*, 12, 76.

—— (1962), *ibid.*, 13, 333.

KINZER, G. D., and GUNN, R. (1951), *J. Met.*, 8, 71.

KOBAYASHI, T. (1955), *Contrib. Inst. Low Temp. Sci., Hokkaido Univ.*, 8, 75.

KOENIG, L. R. (1963), *J. Atmos. Sci.*, 20, 29.

KOMABAYASI, M., and IKEBE, Y. (1961), *J. Met. Soc. Japan*, 39, No. 2, p. 82.

KRASIKOV, P. N. (1961), *Trudy Glavnoi Geofiz. Obs.*, 104, 3.

KUMAI, M. (1961), *J. Met.*, 18, 139.

LANGER, G., and ROSINSKI, J. (1962), Final Report, Armour Res. Foundation, Chicago (ASTIA Doc. No. AD283531, U.S. Dept. of Commerce).

LANGMUIR, I. (1944), *Collected Works*, ed. C. G. SUITS and H. E. WAY (New York: Pergamon Press, 1961), 10, 227–28.

McCLAIN, E. P., and OMOTO, Y. (1962), Univ. Chicago Dept. Geophys. Sci., Sci. Rep. No. 8 (AFCRL-62-849, U.S. Dept. of Commerce).

McDONALD, J. E. (1963), *Zeitschr. angew. Math. Phys.*, 14, 610.

MARSHALL, J. S., and LANGLEBEN, M. P. (1954), *J. Met.*, 11, 104.

MASON, B. J. (1952), *Quart. J. Roy. Met. Soc.*, 78, 377.

—— (1957), *The Physics of Clouds* (New York: Oxford University Press), p. 107.

MORDY, W. (1959), *Tellus*, 11, 16.

MORRIS, T. R. (1957), *J. Met.*, 14, 281.

NAKAYA, U. (1954), *Snow Crystals* (Cambridge, Mass.: Harvard University Press).

NEIBURGER, M., and CHIEN, C. W. (1960), in *Physics of Precipitation* (Amer. Geophys. U. Monogr. No. 5), p. 191.

POWER, B. A., and POWER, R. F. (1962), *Nature*, 194, 1170.

REYNOLDS, S. E. (1952), *J. Met.*, 9, 36.

REYNOLDS, S. E., and BRAHAM, R. R., JR. (1952), *Bull. Amer. Met. Soc.*, 33, 123.

REYNOLDS, S. E., HUME, W., VONNEGUT, B., and SCHAEFER, V. J. (1951), *Bull. Amer. Met. Soc.*, 32, 47.

ROOTH, C. (1957), *Tellus*, 9, 372.

SCHAEFER, V. J. (1946), *Science*, 104, 457.

—— (1949), Progress Reports, Project Cirrus, Gen. Elec. Co., Schenectady, N.Y.

SMITH, E. J., and HEFFERNAN, K. J. (1956), *Quart. J. Roy. Met. Soc.*, 82, 301.

SMITH, E. J., HEFFERNAN, K. J., and SEELY, B. K. (1955), *J. Met.*, 12, 379.

SMITH, E. J., HEFFERNAN, K. J., and THOMPSON, W. J. (1958), *Quart. J. Roy. Met. Soc.*, 84, 162.

SMITHSONIAN INSTITUTION (1951), *Smithsonian Meteorological Tables* (6th ed.; Washington, D.C.

TOMPKINS, L. M., MUUS, D. A., and PEARSON, T. (1963), *J. Geophys. Res.*, **68**, 3537.

VONNEGUT, B. (1947), *J. Appl. Phys.*, **18**, 593.

VONNEGUT, B., and NEUBAUER, R. (1951), *Bull. Amer. Met. Soc.*, **32**, 356.

WEICKMANN, H. (1945), *Beitr. Phys. freien Atmos.*, **28**, 12.

WORLD METEOROLOGICAL ORGANIZATION (1956), *Internat. Cloud Atlas*, **1**, 66.

6

CLOUD-DROPLET
SPECTRA AND GROWTH
BY COALESCENCE

Under various conditions clouds exhibit a wide range of liquid-water contents and distributions of the water among the different-sized droplets. In addition to the spectrum of sizes produced by the growth of a population of droplets from the basic population of nuclei, as discussed in chapter 5, a broadening of the spectrum toward larger droplets is accomplished as a result of collision and coalescence. The different settling rates or terminal velocities of the various sizes cause the collisions to occur.

The warm rain process, that is, the production of precipitation without the aid of ice crystals as in the Bergeron process, can be accomplished when the spectrum becomes broad enough to favor collisions in the gravitational-viscous field.

Amount and Spectral Distribution of Water

Some thirty different groups of cloud physicists have published in recent years results of measurements of drop-size distributions or liquid-water contents, or both, in various types of clouds and fogs. In general, the measurements give the expected result, that as clouds approach precipitation categories a broadening spectrum extending into the large sizes is found. Of course, when rain is actually occurring the size distributions run from tiny cloud droplets to raindrops; however, the greatest interest is in those broad distributions within subprecipitating sizes, for then the possibility of collision-coalescence as a rain-initiating mechanism can be assessed.

From the large amount of published data it is difficult to select material that will summarize conditions for various types of clouds. Many of the measurements are taken with crude instruments, are perhaps unrepresentative, or give ambiguous references as to type of cloud.

It is convenient to consider the two main types of non-precipitating all-liquid clouds—stratus and cumulus. These two kinds of clouds represent opposite extremes of vertical air motions and condensation rates, the stratus being characterized by negligible vertical motions and even distributions of water

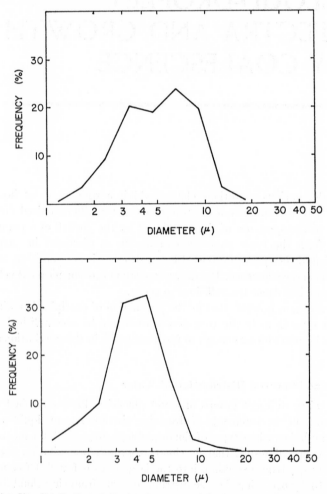

FIG. 6.1.—Size distribution of droplets in non-precipitating stratus cloud as measured by Pedersen and Todsen (1960) at 500 m above sea level in Norway: *upper*, wind less than 4 knots; *lower*, wind greater than 4 knots.

142

while the cumulus clouds are marked by strong vertical currents, intense condensation, and inhomogeneities in liquid-water content.

Distributions in non-precipitating stratus and fogs as measured near Oslo, Norway, by Pedersen and Todsen (1960) are shown in Figures 6.1 and 6.2. The stratus cases are separated into those with wind speeds less than 4 kts and 4 kts or greater. The latter show nearly four times the number concentrations as the former, although the volume mean diameters are nearly the same—9 *vs.* 7 *μ*. The liquid-water contents were not measured independently. Some values for stratus clouds measured by others include measurements of 0.5 g m⁻³ in Cali-

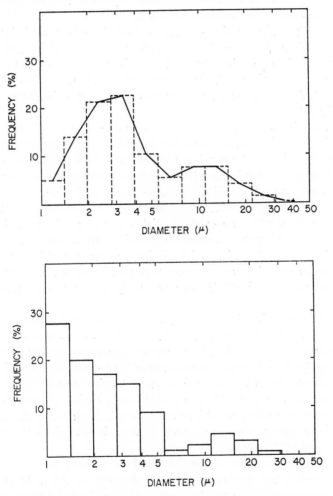

FIG. 6.2.—Size distribution of droplets in fog as measured by Pedersen and Todsen (1960) in Norway: *upper*, advection fog; *lower*, radiation fog (middle of layer).

143

fornia stratus (Neiburger, 1949), 0.6 g m^{-3} in Australia (Warner, 1955), 0.22 g m^{-3} in the USSR (Nikandra and Khimach, 1960), and 0.65 g m^{-3} in thick stratus in Arizona (Byers, 1959) with a maximum point at 1.5 g m^{-3}.

Singleton and Smith (1960) reported on measurements in layer clouds of different thicknesses in the vicinity of the British Isles. In a layer 700 ft thick they found 0.20 g m^{-3} liquid-water content in droplet concentrations of 316 cm^{-3}; in two measurements of 1000-ft-thick clouds they measured 0.78 g m^{-3} and 1.88 g m^{-3}, the former on 392 droplets per cm^3 and the latter on 225 cm^{-3}. The greatest liquid-water content was 3.15 g m^{-3} with 106 droplets per cm^3 in a layer 3500 ft thick. As might be expected, they found that the size spectrum becomes broader as the thickness increases.

The most intensive studies of cumulus clouds have been made by the University of Chicago group (Braham *et al.*, 1957; Battan and Reitan, 1957). Mean distributions of droplet sizes obtained in trade-wind cumulus and in cumulus clouds over the continental United States in the summer are shown in Figures 6.3 and 6.4. The echo-producing clouds shown in the figures in all cases produced very light rain.

Ackerman (1959) analyzed the data for the vertical distribution of water content in the trade-wind cumulus. She found that the quantities were highest in the upper third of the cloud and lowest near the bottom. The clouds extended from about 2000 feet to a height of about 6000 feet. When the average of each traverse was taken and all traverses at all heights were combined, the mean of these averages was 0.54 g m^{-3}, and when the maximum of each traverse was combined for all traverses at all heights a mean maximum of 1.50 g m^{-3} was found. Draginis (1958) examined the maxima and concluded that the water content is less than that of an adiabatically ascending parcel in non–echo-type clouds and exceeds that value in echo-type clouds. He also noted that clouds with a maximum less than 1.75 g m^{-3} invariably dissipated before producing an echo.

Although the University of Chicago flights show local regions in cumulus clouds where the liquid-water content exceeds that which would be achieved by closed parcel ascent, it is found that the average values are considerably less. Ackerman (1963) took cumulus-cloud data of Warner (1955) and Squires (1958) and compared them with Chicago data on trade-wind cumulus and hurricanes. When the data were plotted in terms of the ratio of observed to parcel-ascent water content a distribution with height as seen in Figure 6.5 is shown. Only in the hurricane flights, which were made in the rain bands, did the convective clouds extend to the upper flight levels.

By means of a special device for counting large droplets, Brown and Braham (1959) extended the spectral range to a diameter of 300 μ. In continental cumulus congestus they found that larger clouds which build up to higher altitudes have higher number concentrations in the range of 75 to 300 μ diameter

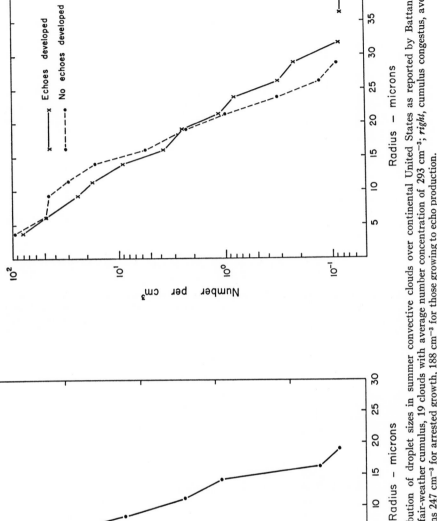

Fig. 6.3.—Distribution of droplet sizes in summer convective clouds over continental United States as reported by Battan and Reitan (1957): *left*, fair-weather cumulus, 19 clouds with average number concentration of 293 cm⁻³; *right*, cumulus congestus, average droplet concentrations 247 cm⁻³ for arrested growth, 188 cm⁻³ for those growing to echo production.

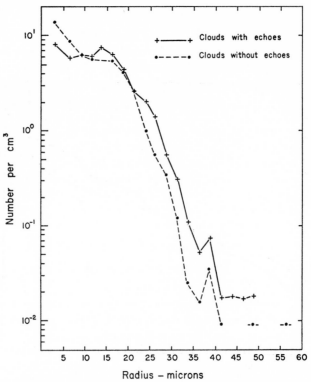

Fig. 6.4.—Distribution of droplet sizes in trade-wind cumulus over the ocean near Puerto Rico as reported by the Chicago measurements (Braham *et al.*, 1957). In 11 clouds which grew to produce radar echoes, the number concentrations were 52 cm⁻³, and in 26 non-echo clouds they were 58 cm⁻³.

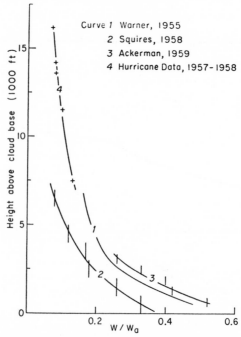

Fig. 6.5.—Ratio of measured liquid-water content to that produced by parcel ascent, plotted for various cloud situations by Ackerman (1963) in terms of height above cloud base. Curves *1* and *2* are for clouds near Australia, *3* for trade-wind cumulus in the Caribbean, and *4* for West Indian hurricanes.

than do small clouds. For all drops collected in this size range, the average concentration was 1492 per cubic meter with a maximum of 7581.

Methods of Measurement

There are a variety of ways in which droplets can be captured for counting and sizing. In the University of Chicago flights the "shotgun" sampler (Brown and Willett, 1955), named for the resemblance of its action to that of a shotgun, was used successfully. Through a tube extending out from the side of the airplane, three glass slides coated with silicone oil are forced one at a time across the air stream into a storage tube beyond. After the third one is exposed, the gap between the storage end and the main tube is closed and the slides, with their mountings now coupled together, are pulled through the tube into the airplane where they are immediately photographed under the microscope in a cold cabin. The collection efficiency of the slide at the various air speeds flown is calibrated in a water-spray wind tunnel.

The sparsely distributed droplets of diameters significantly greater than 50 μ are mostly missed by the shotgun sampler. Brown (1961) perfected a foil sampler which successfully recorded the larger sizes while showing nothing of the smaller sizes. A continuous tape of lead foil backed by a fine-mesh copper-wire cloth is driven by an electric motor past a slit exposed to the air stream. Each drop above a certain size at the air speeds flown makes a dent in the foil as it hits. The sizes of the dents are calibrated for drop sizes.

Two types of liquid–water-content meters are used on the University of Chicago flights to assure coverage in as wide a range of droplet sizes as possible. For the smaller sizes the hot-wire instrument is most successful. Based on a design by Neel and Steinmetz (1952, 1955), this instrument is now available commercially. Droplets strike a heated, looped wire suspended across the air stream in a short tube of $1\frac{1}{2}$ to 2 in. diameter. The sensible heat and heat of vaporization of the droplets that strike the wire modify its temperature. Mounted with this wire, but along the air flow so that it cannot intercept droplets, is a similarly heated wire. The ratio of energy exchange from the two different wires is calibrated in terms of liquid-water content.

For droplets in the peak range of cloud water the paper-tape meter is used. This instrument was originally designed by Warner and Newnham (1952) in Australia. It is the device adapted by Brown for the foil sampler described in a preceding paragraph. A continuous paper tape is driven by an electric motor past a slit where it becomes wetted by the cloud water. There the electrical conductivity, which is proportional to the wetness, is measured from a contacting wire. The liquid-water content is calibrated according to the speed of the airplane, the speed of the tape, the width of the opening, and the electrical conductivity across the paper.[1]

[1] For a more complete description of these and other instruments used in cloud physics see Barrett (1960).

Anyone who has flown through clouds, particularly convective ones, has noted that the density of the clouds as indicated by the visibility along the wing is highly variable. This could be due to a change in the liquid-water content or a change in the droplet sizes or both. Take, for example, droplets of a radius of 10 μ. If the liquid-water content were doubled, the number of droplets of this size necessary to provide the water would be doubled. If the radius were doubled without changing the liquid-water content, the number concentration would be reduced by a factor of eight. The optical effects of number concentrations versus size and total mass are complicated and will not be treated here, but it is interesting to note in passing that as droplets grow the visibility in the cloud may be improved even when the liquid-water content is increasing slightly.

Terminal Velocity and Collision-Coalescence

The concept of terminal velocity developed in chapter 4 for aerosol particles is applicable also to water and ice spheres. For water droplets the Stokes law regime prevails at radii less than 40 μ, with the terminal velocity given from (4.12) by

$$U_T = \tfrac{2}{9} \frac{\rho_L - \rho_a}{\eta} g \, r^2 . \tag{6.1}$$

For larger sizes the terminal velocity is usually taken from values determined experimentally. In Figure 6.6, where terminal velocities through two ranges of sizes are given, the velocities for $r > 40 \mu$ are those obtained by Gunn and Kinzer (1949) by an experiment described briefly in chapter 5.

The most obvious case of collision and coalescence is that in which a raindrop or drizzle drop with an appreciable rate of fall—let us say, $U_T > 20$ cm sec^{-1} and therefore $r > 40 \mu$—collides with droplets of 5 to 10 μ radius which have terminal velocities one or two orders of magnitude slower and hence may be considered as relatively stationary.

A raindrop of radius R falling at terminal velocity U_T sweeps out a cylinder of volume $\pi R^2 U_T$ per second. If the air space contains small droplets of various radii r_i, the mass of water in these droplets per unit volume is $\tfrac{4}{3}\pi \rho_L \Sigma_i N_i r_i^3 = \chi$, which we have called the liquid-water content. If the raindrop collides with droplets representing a fraction E_1 of the water in its path—some of the droplets may be carried around the drop in the air stream—and if a fraction E_2 of the colliding water sticks to the drop, the mass of water collected per unit time would be

$$dM/dt = \pi R^2 U_T \chi E_1 E_2 , \tag{6.2}$$

and, since $M = \tfrac{4}{3}\pi R^3 \rho_L$,

$$\frac{dR}{dt} = \frac{dM/dt}{dM/dR} = \frac{U_T \chi E_1 E_2}{4 \rho_L} . \tag{6.3}$$

Terms E_1 and E_2 are called the collision efficiency and coalescense efficiency, respectively. Their product is referred to as the collection efficiency. The relation implies zero terminal velocity for the collected droplets. If the droplets other than the collector were all of the same size, therefore of uniform terminal velocity u_T, the velocity factor in (6.3) would be $U_T - u_T$. In the following discussion, except where otherwise noted, the single large drop will be considered as moving among droplets whose terminal velocity is zero.

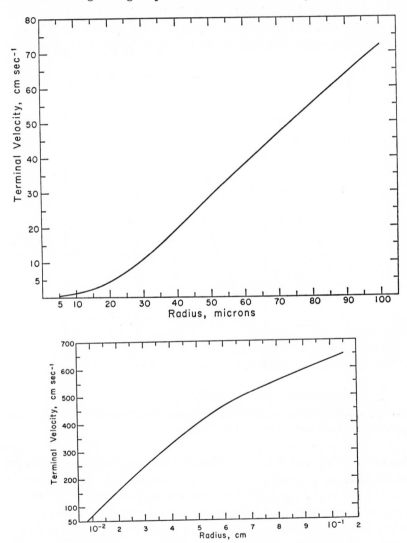

Fig. 6.6.—Terminal velocity of spheres of unit density as a function of their radii in two different size ranges.

149

The determination of the coalescence efficiency E_2 is extremely difficult, since it depends on such factors as surface chemistry and electrostatics. Justifiably or not, it has frequently been taken as unity in various computations. We shall consider it after first examining methods of determining the collision efficiency E_1. The latter can be approached through the analysis of the streamlines around a collecting sphere and the motion of the small droplets with respect to these streamlines.

In the analytical and experimental determination of the air flow around and in the vicinity of a sphere the Reynolds number Re must be taken into account. As already defined in equation (4.15), $Re = 2\rho_a RU/\eta$, where R is the

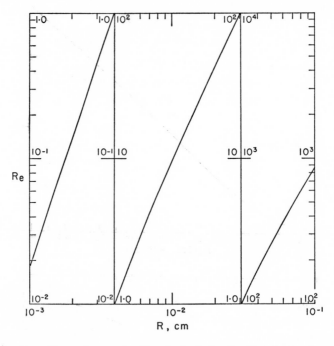

Fig. 6.7.—Reynolds number as a function of spherical radius. Note that scale runs in three ranges of Reynolds numbers: 10^{-2} to 1.0, 1.0 to 10^2, and 10^2 to 10^4.

radius of the sphere and U is the undisturbed air velocity in its vicinity. For an object falling at terminal velocity, $U = U_T$ and, since the terminal velocity of a sphere in air of given density is a function of R and the drag coefficient, the Reynolds number for flow about a sphere at a given temperature and pressure is a function of the radius only. The graph in Figure 6.7 shows Re as a function of R at 900 mb and 0° C.

Different Treatments for Different Reynolds Numbers

The streamlines around the spherical collector or the motion of the two spheres (collecting and collected) relative to each other have received different analytical treatments depending on the Reynolds number and, to some extent, the choice of the investigators. Four different ranges of Reynolds numbers have called forth different approaches, as follows:

(1) For $Re > 100$ corresponding to $R > ca.$ 300 μ, potential air flow is approximated, i.e., the streamlines are considered as evenly spaced and concentric in incompressible air flowing around the sphere ($c = -\nabla\varphi$, where c is the velocity and φ is the velocity potential, and $\nabla^2\varphi = 0$). This is the method introduced by Langmuir and Blodgett (1946) with later refinements by Langmuir (1948), Ludlam (1951), and Mason (1957).

(2) For $0.5 < Re < 20$, approximately, corresponding to $19 \mu < R < 140 \mu$, disturbance of the flow field by the presence of both spheres (drop and droplet) is considered. This range was investigated by Shafrir and Neiburger (1963).

(3) For $20 < Re < 100$, corresponding to $140 \mu < R < 300 \mu$, an interpolation between values was suggested by Langmuir (1948).

(4) For $Re < 0.5$, corresponding to $R < 20 \mu$, the Stokes approximation holds for both spheres, which in this case are near the same size. This range was studied mainly by Hocking (1959).

Potential Flow

The assumption of potential flow, which bears a close semblance to reality for raindrops of appreciable size, is the simplest case to deal with and the only one that will be shown in detail here. The small droplet moves through the streamlines because of its own inertia but does so with a Stokes law velocity. The result is a path with respect to the collecting sphere intermediate between inertial impact and streamline flow, as illustrated in Figure 6.8.

In the enlarged representation, OA is the airflow vector tangent to a streamline at O, and OB is the vector for the particle velocity at O. The particle, however, follows a continually changing path to C (upper sketch). It moves relative to the streamlines with a Stokes law velocity $U - u$, represented for the point O by the vector AB, but for other points along the streamlines by other vectors of $U - u$.

In accordance with a well-known practice of fluid dynamics, non-dimensional parameters are introduced in order to simplify the model. One then ends up with a coefficient involving only the undisturbed flow U_0—actually the terminal velocity of the sphere U_T or, if the droplet also has an appreciable velocity, $U_T - u_T$—and the radii R and r. The non-dimensional parameters are

$$\varphi = x/R, \qquad \Xi_x = U_x/U_0,$$
$$\psi = y/R, \qquad\qquad\qquad (6.4a)$$
$$\tau = U_0 t/R, \qquad \Xi_y = U_y/U_0.$$

151

Then

$$\frac{dx}{dt} = \frac{U_0 d\varphi}{d\tau}, \qquad \frac{d^2x}{dt^2} = \frac{U_0^2}{R}\frac{d^2\varphi}{d\tau^2}, \qquad (6.4b)$$

with a similar expression for dy/dt, etc.

For the droplet, in accordance with the Stokes law flow, the inertial force balances the viscous force or

$$M_i \frac{du}{dt} = 6\pi\eta r_i |U - u|, \qquad (6.5)$$

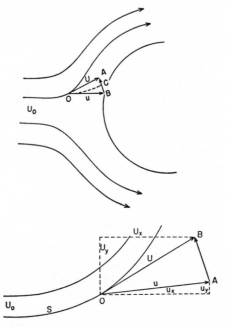

FIG. 6.8.—Geometry of potential flow, enlarged in lower diagram

where the subscript i refers to the droplet and $|U - u|$ is a vector. Substituting density times volume for M_i, we have

$$\frac{du}{dt} = \frac{9}{2}\frac{\eta}{\rho_L r_i^2}|U - u|. \qquad (6.6)$$

Expressed in terms of coordinates of x and y of the cross-section of the sphere,

$$\frac{d^2x}{dt^2} - \frac{9}{2}\frac{\eta}{\rho_L r_i^2}(U_x - u_x) = 0, \qquad (6.7)$$

with a similar expression in y. Rearranging, we have

$$\frac{2}{9}\frac{\rho_L}{\eta}r_i^2\frac{d^2x}{dt^2} + \frac{dx}{dt} - U_x = 0. \qquad (6.8)$$

152

The quantities substituted from equations (6.4a, b) give

$$\frac{2}{9}\frac{\rho_L}{\eta}\,r_i{}^2\,\frac{U_T{}^2}{R}\frac{d^2\varphi}{d\tau^2}+U_T\frac{d\varphi}{d\tau}-U_T\Xi_x=0\,,\qquad(6.9)$$

with a similar expression in terms of y. Then taking out the common factor U_T and letting

$$\frac{2\rho_L r_i{}^2 U_T}{9\,\eta R}=2K\,,\qquad(6.10)$$

we find

$$2K\frac{d^2\varphi}{d\tau^2}+\frac{d\varphi}{d\tau}-\Xi_x=0\,,\qquad(6.11)$$

and in terms of y a similar expression.[2]

The nature of K is apparent when we note that

$$K=\frac{\Xi_x-d\varphi/d\tau}{2d^2\varphi/d\tau^2}=\frac{1}{9}\frac{\rho_L}{\eta}\frac{r_i{}^2}{R}U_T=\frac{(U-u)/2R}{\dot u/U_T}\,,\qquad(6.12)$$

where the subscript x is dropped from the velocities and $\dot u=du/dt$. The K as defined in the second equality in (6.12) is a non-dimensional factor dependent only on the radii of the drop and droplets and the viscosity of the air. The terminal velocity U_T is also a function of R. Except for snow pellets and some forms of hail, ρ_L may be taken as 1. The value of η is mainly a function of altitude, but is assumed constant in earlier work on the subject. Physically, from the last equality in (6.12), K is seen to be the deviation in velocity of the droplet from that of a streamline in the distance $2R$ (one diameter) divided by the deceleration per unit of initial relative velocity. In other words, K is a measure of the inertia of the droplet as modified in a viscous fluid.

For the collision efficiency in potential flow, E_p, Langmuir (1948) found the relationship

$$E_p=\left(\frac{K}{K+\frac{1}{2}}\right)^2\qquad(6.13)$$

was valid for values of $K>0.2$.

For viscous flow, Langmuir arrived at another relationship

$$E_v=\left[\frac{K-1.214}{1+(0.75\ln 2K)}\right]^2.\qquad(6.14)$$

He showed experimentally that the conditions could be met by a general expression

$$E_1=\frac{E_v+E_p Re/60}{1+Re/60}\,,\qquad(6.15)$$

[2] The factor 2 enters because, following the original derivation, diameter is considered rather than radius. In terms of diameters d and D, it turns out that K is given by $\rho_L d_i{}^2 U_T/18\eta D=\rho_L r_i{}^2 U_T/9\eta R$. The K without the factor 2 has priority, although in some ways $K'\equiv 2K$ is simpler to work with.

153

such that $E_1 \cong E_v$ when Re is small and $E_1 \cong E_p$ when Re is much larger than 60, with transitional values in between. The value of Re is 60 for a radius of about 225 μ.

The viscous case has been treated in a different way by other investigators, as will be seen presently.

Ludlam (1951) modified the Langmuir approach by taking into account the fact that collision would occur on a streamline that carries the center of a droplet within a distance r of the perimeter of a collector; Langmuir had considered the smaller droplets as points, with grazing incidence only on streamlines through their centers.

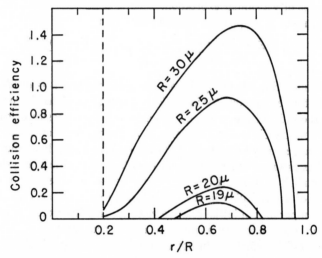

Fɪɢ. 6.9.—Collision efficiency according to Hocking (1959) for drops of radius R and droplets of radius r.

Pairs of Spheres in Viscous Flow

As the collecting sphere is made smaller in the computations, it approaches in size the droplets to be collected. It then becomes necessary to treat pairs of spheres in viscous flow. The Stokes approximation can be applied to the hydrodynamical equations describing the motion of two neighboring spheres at low Reynolds numbers, taking account of the mutual interference of the flows.

Hocking (1959) found a solution for small drops ($Re < 0.05$) by computing the drag force due to the flow field around both spheres. The accuracy decreases when the spheres are of very different sizes, so the method is not applied if the larger drop is more than five times the size of the smaller. But if the sizes are too similar—within 10 to 20 per cent—the spheres repel each other.

Hocking's results for collecting drops of radii 19 μ, 20 μ, 25 μ, and 30 μ are reproduced in Figure 6.9, where collision efficiency is plotted against ratio of droplet radius to collector radius. The values show a wide variation quite sensitive to changes in the radii. Efficiencies exceeding 1.0 are seen to be possible in some cases, indicating a "capture" effect. The low collision efficiencies of the smaller sizes are strikingly shown. Considering that a droplet of radius 20 μ is large as sizes go in non-precipitating clouds, one can understand the stability of most clouds.

For Reynolds numbers larger than those considered by Hocking, up to about 20, Shafrir and Neiburger (1963) approximated the drag force on each sphere using only the disturbance of the flow field due to the other. The final computation requires solution of the stream-function field and the vorticity field. The problem is developed from a treatment to be found in Lamb's *Hydrodynamics* (1932), following Oseen (1910).

The Results and Their Meaning

Mason (1957) has put together the results of various studies in the drop-spectral regions of their best applicability. Using his tabulation and the more recent computations of Hocking and of Shafrir and Neiburger at Reynolds numbers less than 20, one obtains values for collision efficiencies such as those represented in Figure 6.10. Current work is likely to change the values to some extent, so it is considered unwise to present complete tables here.

In this and subsequent discussions we refer to the collector as the *drops* and the collected particles as the *droplets*. A typical stable cloud with droplets of 6 μ radius would, as seen in Figure 6.10, have a collision efficiency of 0.60 to 0.75 on drops of radii greater than 200 μ (0.2 mm). Through the range of drops of drizzle size the efficiency changes rapidly with size. For droplets of 10 μ radius the collision efficiency changes very little from the smallest drizzle drops (radius about 50 μ) to the largest raindrops, varying between about 0.8 and 0.9 through this range. The dropping off of the efficiency as Hocking's range is entered is most striking.

For droplets of radii 15 μ and 20 μ the "capture" effect at small sizes is noted. At raindrop sizes the collision efficiency hovers close to the 1.0 value.

These collision efficiencies are in agreement with the concept already expressed that after precipitating sizes are reached collision-coalescence becomes the dominant growth process. For collectors in the range of radii between 20 and 40 microns, interest centers on the possibility of *initiation* of rain by an all-liquid process. If Hocking's collision efficiencies are truly representative, it is then evident that as they grow from a radius of 20 μ to one of 30 μ the drops become good collectors with collision efficiencies exceeding 1, provided the collected droplets have radii of about 10 μ or larger. Examples of growth times computed in 10-micron steps are given in Table 6.1. It is seen that after the collecting drop has reached a radius of 30 μ the times for growth to larger sizes become reasonable. The table is computed for a liquid-water content of 1 g m^{-3}.

155

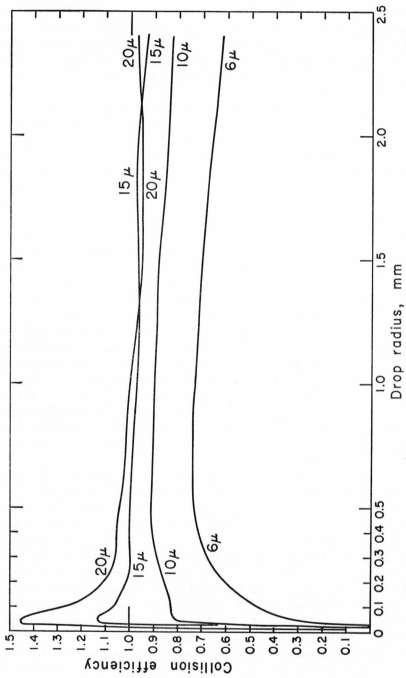

FIG. 6.10.—Collision efficiencies supplied from sources given in text. Abscissa is radius of collecting drop and curves are for various radii of collected droplets.

The times given should be divided by the liquid-water content for values other than 1 g m^{-3}.

Distributions of sizes in natural clouds, interpreted in terms of these collision efficiencies, suggest the possibility of rain formation by this process in warm clouds, if it is assumed that the coalescence efficiency, E_2, is near unity. Observations seem to bear out the reasonableness of this assumption. But observations also show that it is only in developing cumulus clouds that drops with radii of 30 μ or more are found in reasonable numbers or that liquid-water contents are as high as 1 g m^{-3}. In this connection it should be remembered that giant nuclei such as NaCl may attain this size at 99 per cent relative humidity.

TABLE 6.1

TIME REQUIRED FOR GROWTH OF A DROP OF INITIAL RADIUS R_1 IN A CLOUD OF
DROPLETS OF RADIUS r CONTAINING 1 g m^{-3} LIQUID WATER

(Computation Made in Steps)

Increase in Radius (μ)	$r=10\ \mu$ $R_1=20\ \mu$	$r=15\ \mu$ $R_1=20\ \mu$	Increase in Radius (μ)	$r=6\ \mu$ $R_1=30\ \mu$	$r=22\ \mu*$ $R_1=30\ \mu*$
20–25.........	52′54″	35′16″	30–35.........	52′54″	1′47″
25–35.........	8′43″	4′54″	35–45.........	11′30″	2′47″
35–45.........	4′26″	3′0″	45–55.........	7′11″	1′36″
45–55.........	2′42″	2′2″	55–65.........	4′52″	1′14″
55–65.........	2′8″	1′33″	65–75.........	3′34″	1′1″
65–75.........	1′43″	1′16″	75–85.........	2′52″	0′53″
75–85.........	1′23″	1′6″	85–95.........	2′19″	0′48″
85–95.........	1′16″	0′58″			

* The combination $R_1 = 30\ \mu$, $r = 22\ \mu$ gives the highest collision efficiency from Hocking's data.

Experimental Evidence

A number of experimental measurements of collision efficiencies, coalescence efficiencies, or the combined collection efficiencies have been made in the decade of the 1950's. While the widely differing approaches sometimes have presented conflicting results, a reasonable picture can be pieced together from most of them.

Gunn and Hitschfeld (1951) measured the gain in weight of drops falling in a column 3 meters high. The drops were initially of 1.5 mm radius and the droplets 6 to 100 μ. Good agreement was found with Langmuir's collision efficiencies taken as the true collection efficiencies, in other words, with coalescence assumed in every collision.

Telford, Thorndike, and Bowen (1955) worked with smaller drops, $R = 75\ \mu$, thrown off from a spinning disk. These were thrown into a vertical wind tunnel to produce a cloud of sizes within ± 6 per cent of the same radius. The collisions were carefully photographed. They found collection efficiencies as high as 12.

157

These high values came about by a common occurrence in which an upper drop was captured in the wake of a lower one.

Kinzer and Cobb (1958) observed single drops falling through a dense cloud of 1500 to 3000 droplets per cm³ having radii of 5.5 to 8 μ. The measurements were made in a vertical wind tunnel. The collection efficiency on drops of 8 μ radius confirmed Langmuir's calculations for the viscous case, but for a radius of 20 to 40 μ the collection efficiency dropped to 0.2. Then it was near 1 for a radius of 200 μ, but fell to low values at a drop radius of 1000 μ.

Schotland (1957) experimented with a model consisting of steel balls in a sugar solution, scaling his model to an Re of atmospheric conditions. He noted the wake effect of capture of one sphere in another's wake.

Electrical Effects

Doubt is cast on some of the experimental determinations because the electrical effects were unknown or uncontrolled. Electrical effects are shown to be of great importance, but the available data do not present a unified picture.

Levin (1954) demonstrated theoretically that an electric charge multiplies the collection efficiency of small collectors, especially those having a radius less than about 15 μ. Kinzer and Cobb (1958) considering cases where both the collector and the collected droplets were charged found little effect for radii greater than 8 μ.

Telford *et al.* (1955) found that unlike charges of 6×10^{-4} e.s.u. on drops of radius 65 μ increased the collection rate 2 to 200 times. This is a much higher charge than that found in natural clouds.

Twomey (1956) measured charges on droplets in natural clouds by observing the motion of the droplets in an electric field. He found the charges amounted to $Q = 3 \times 10^{-9} \, r^2$ e.s.u. (r in microns). In the clouds, 50 per cent of the droplets were charged, and of these 80 per cent were positive. But Phillips and Kinzer (1958) found in non-precipitating clouds on a mountain peak that positive and negative charges were about equal in numbers. For droplets 4 μ in radius they found only 2.4×10^{-9} e.s.u., or one-twentieth the value to be expected from Twomey's relation. In thunderstorm clouds, however, they found charges two orders of magnitude larger with either positive or negative predominating, though sometimes in about equal numbers.

A number of measurements of electrical properties of clouds and cloud particles have been made in the Soviet Union (Krasnogorskaya, 1963) from airplanes and mountain stations. Charge on cloud droplets was measured from their displacement in passing through an electric field as in the case of Twomey's measurements, while the charge on larger raindrops was obtained from their effects upon passage through an induction ring. Raindrop sizes were measured by a photoelectric method. Electric fields were measured with well-known field-mill methods. As in the observations of Phillips and Kinzer the data show almost equal frequency of negatively and positively charged drops, but in clouds as well as precipitation considerable areas of particles of predominantly the same sign were noted.

On droplets of radius 1 to 14 μ Krasnogorskaya found charges ranging between about $\pm 6 \times 10^{-8}$ e.s.u. distributed approximately normally about a peak near zero and depending on droplet size. In warm cumuli the average drop charge was related to size in a linear fashion according to $\bar{q} = kd$, where d is diameter in cm and k has a value of 4×10^{-5} e.s.u./cm. Raindrops and snow had charges ranging roughly from 10^{-4} to 10^{-2} e.s.u., the larger values being found in the Caucasus mountains near Mt. Elbrus at 3080 m in shower clouds, the smaller values at 2140 m in continuous rain.

From theoretical computations Krasnogorskaya finds that drops of 10 μ radius in a cloud of droplets of 6 μ radius carrying charges of $+10^{-6}$ and -10^{-7} e.s.u., respectively, would have a collision efficiency of 0.31, whereas in the absence of charge the efficiency would be zero. The effect of an electric field is shown from the computations considering the same drops and droplets uncharged. With an electric field of 1200 volts per cm, the collision efficiency of the uncharged particles becomes 1.94. The more nearly the same size the droplets may be, the greater the enhancing effect of the electric field. For example, if the above-mentioned uncharged collected droplets were of radius 8 μ instead of 6 μ the collision efficiency at 1200 volts/cm would go to 4.59. Electric fields of this order of magnitude are found in thunderstorms, and if these theoretical computations represent the conditions in nature a great enhancement of drop coagulation would be expected in thunderstorms, especially if the additional enhancement in the case of drops having opposite charges is considered.

Goyer *et al.* (1960) have reported on experiments on *coalescence* efficiencies of drops in electric fields of various strengths. They produced primary drops of 300 to 395 μ radius interspersed with smaller ones of about 50 μ radius. The drops were formed and set in motion in a jet created by forcing water through a capillary downward into an open space between two horizontally placed condenser plates. The water jet broke into drops very quickly after ejection. As a result of a sudden decrease in surface energy and increase in kinetic energy accompanying the breakup, the small satellite droplets moved faster than the large primary ones with the result that they frequently collided with the drops ahead. The collisions, occurring about 1 to 2 cm below the capillary tip, were photographed by a high-speed motion-picture camera (4000 to 7000 frames per second), and the ratio of coalescence to collisions was obtained by examining the film. Various electrical potentials were applied between the plates.

It was found that the coalescence efficiency was greatly enhanced by the imposed electrical fields. A repulsion effect resulting from a net charge of like sign on the drops was noted when the field reached values in excess of several hundred volts per cm. The coalescence efficiencies as a function of field strength are given in Table 6.2.

159

The electrical effects have great practical meaning for natural clouds. While colloidally stable clouds exhibit neither significant charges on the droplets nor measurable electric fields, the stage in a cloud's development when raindrops begin to grow is marked by an increase of the electric field to some 10 volts per cm and the development of an appreciable charge on the drops. The collection efficiency must then increase, with the coalescence component, E_2, becoming larger than 1.

TABLE 6.2

COALESCENCE EFFICIENCIES AS A FUNCTION OF
FIELD STRENGTH IN COLLISIONS BETWEEN
DROPS OF 300 TO 395 MICRONS IN RADIUS
AND DROPS *ca.* 50 MICRONS IN RADIUS

Field Strength (volts/cm)	Coalescence Efficiency (per cent)	Standard Deviation
0.0	29.4	3.1
3.1	33.5	2.4
15.4	88.7	3.6
38.4	95.3	1.6
923.0	0.0	

Observations of Rain in Warm Clouds

As was pointed out in chapter 5 the formation of rain by the collision-coalescence process is more common than was formerly thought. Summer cumulus clouds in the eastern and southern United States, for example, have been shown by the University of Chicago observations to initiate rain normally without the presence of the ice phase.

Langmuir (1948) has used the expression "chain reaction" to describe the development of rain in warm clouds. Drops grow by coalescence until they reach a certain maximum size above which they cannot be maintained by cohesion and must break up into several drops each. These, in turn, grow in the same way, such that if each drop breaks up into n pieces, each of which in turn grows to breakup size and so on through p cycles, one drop produces n^p drops.

Studies by radar of the sequence of events following first appearance of echoes in cumulus clouds have been made by the Chicago group at the Missouri field site. A typical sequence is for the echo to develop downward faster than the fall speed of the raindrops, even if consideration is given to a helping downdraft. Below the cloud base the echo descends at the rate to be expected from falling rain, showing further that there is no downdraft effect of importance. The fast descent inside the cloud can only be interpreted as a downward propagation of the rain-producing process. No clear picture of the microphysics of this rapid downward spread of the echo is apparent. The same phenomenon appears in the

160

radar studies of Battan (1953) from the Thunderstorm Project in Ohio. A cloud electrification effect is suspected.

Fall Speed of Solid Hydrometeors

A number of observations have been made of the densities and fall speeds of snow and ice particles, beginning with work of Nakaya and Terada reported in 1934. They allowed individual snow particles to fall through a vertical cylinder 2 m in height, closed at each end to eliminate air motions, and observed the time of fall. After passing through the cylinder each was caught on a glass slide coated with paraffin, then photographed. Next the ice was melted and the diameter of the resulting approximately hemispherical drop was measured in order to determine its original mass.

A striking finding was that the fall speed of planar snow crystals—plates and dendrites—was the same regardless of size. Needles and rimed crystals had higher terminal velocities the larger the size. Plane dendrites all had terminal velocities of about 30 cm sec^{-1}, "powder" snow 50 cm sec^{-1}, and spatial dendrites 55 to 60 cm sec^{-1}. Needles 0.5 mm long fell at about 25 cm sec^{-1}, 1 mm at 50 cm sec^{-1}, and 2 mm at about 70 cm sec^{-1}. Rimed crystals of about 1.5 mm had terminal velocities of 85 to 90 cm sec^{-1}, while those of 3.3 mm fell at 110 cm sec^{-1}.

The masses, in milligrams, were found to be related to the maximum dimension, d (mm), by the relations

0.0029 d for needles,
0.0038 d^2 for plane dendrites,
0.010 d^2 for "powder" snow and spatial dendrites,
0.027 d^2 for rimed crystals.

Snow or ice pellets, often called by the German name *Graupeln* (sing. *Graupel*), form a class that is of special interest because they are commonly found in summer cumulus just above the freezing level. Nakaya and Terada (1934) caught pellets from winter clouds which had an approximately uniform density of 0.125 g cm^{-3}. Pellets collected by Braham (1963a, 1964) and co-workers, apparently formed from the freezing of drizzle-sized drops in the ascending currents of summer cumulus, characteristically have a density near 0.9 g cm^{-3}. After being collected in an airplane, they were preserved in a cold container, then dropped down a 10-meter tube mounted in a food-freezing plant. The speed attained near the bottom of the fall was determined photographically. The fall speed as a function of size is shown in Figure 6.11, where a comparison is made with the terminal velocities of liquid drops as determined by Gunn and Kinzer (1949) and with the data on the graupels measured by Nakaya and Terada. The spots are sketches drawn to true relative scale of the pellets as they appeared in the photographs. The Nakaya and Terada pellets appear to have

161

been formed by the clumping together and compacting of snow crystals after riming to form a less dense pellet than would be the case with freezing drops.

As explained by Magono (1954), the lack of dependence of the fall velocity of plates upon their size is due to the tendency of all platelike structures to fall with their planes horizontal. If the thickness of the plate is constant, both the mass and the aerodynamic drag are proportional to its area.

In an experiment using photographic methods to determine speed, and dyed

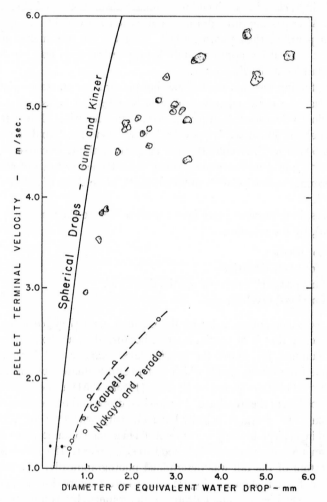

Fig. 6.11.—Terminal velocities of various sizes and shapes of snow pellets as sketched to uniform scale as measured by Braham (1963b). Comparison is made with data on spherical drops and on "graupels" presumably of lower density, by Nakaya and Terada. (Unpublished figure, courtesy of Professor Braham.)

paper to obtain size, Langleben (1954) found the terminal velocity to be given by $u = kd^n$, where d is the diameter in cm of the droplet formed after melting the snowflake. The parameters k and n varied with shape, amount of riming, and melting, but n was usually near 0.3 while k was 160 for dendrites, 234 for combinations of plates and columns, and increased rapidly with riming or melting.

From aerodynamic considerations Magono (1953) found for pristine forms $u = 132\,r^{1/2}/(0.40 + 0.63r)^{1/2}$, where r is the radius in cm of the flake considered as approximately spherical. For rimed flakes his results showed the first factor to be 194 instead of 132 and the other numbers essentially the same. It is apparent that when r is small u depends simply on $r^{1/2}$, but for large flakes the r's in the numerator and denominator nearly cancel each other, and a fall velocity independent of size is approached.

Growth of Snow Flakes by Collision-Coalescence

In the absence of riming or electrostatic effects it is probable that snow crystals seldom coalesce with each other. At low temperatures where even the lowest clouds have turned to ice crystals only the rudimentary pristine forms are found. In their slow descent through the clouds there is time for a considerable amount of growth by diffusion from the vapor. It is common observation that the largest snow flakes occur when the temperature is very near freezing, and under these conditions there is always strong evidence of riming. The reason that riming helps collision and coalescence is two-fold: in the first place rimed particles have different fall speeds depending on size, thus making collisions possible, while umrimed planes tend to have the same rate of fall regardless of size; second, the presence of liquid on the surfaces helps to cement them together.

From these arguments it would appear that needles, which have size-dependent fall speeds, could collide more easily, and this is exactly what is observed. Needles often occur in bunches or sheaths or even in clumps with axes non-parallel. Since the needle represents a growth habit characteristic of temperatures near 0° C, it is often formed in an environment favorable for riming.

Magono (1953) has calculated the rate of growth of snowflakes by accretion of ice crystals, under the assumption that the terminal velocities of both the collected ice crystals and the collecting flakes are independent of size. It is assumed that the crystals add their volume to the volume of the collector without change in density. If there are N crystals per unit volume of space, each with its own volume V_2, then the growth rate of the collector would be

$$dV_1/dt = NV_2EA_1(u_1 - u_2),\qquad (6.16)$$

163

where the subscript 1 refers to the collector and 2 to the collected crystals; E is the collection efficiency. The cross-sectional area of the collector, A_1, if considered as a sphere, may be written as $aV_1^{2/3}$, where $a = (9\pi/16)^{1/3}$. Then

$$V_1^{-2/3}dV_1 = NV_2Ea(u_1 - u_2)dt \,, \tag{6.17}$$

which, upon integration, gives

$$V_1 = V_0 + \tfrac{1}{3}[NV_2Ea(u_1 - u_2)t]^3 \,, \tag{6.18}$$

where V_0 is the initial volume of the flake.

A more important question is that of the accretion of water droplets on a snowflake. In a treatment similar to that of Langmuir and Blodgett, discussed earlier in this chapter, Ranz and Wong (1952) have investigated the impaction

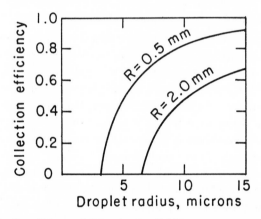

FIG. 6.12.—Collection efficiency of disks of radius R falling among droplets of radius given by abscissa, as computed by Fletcher (1962) from a formula by Ranz and Wong (1952).

of particles on various shapes of collectors. Fletcher (1962) has applied their results to a disk falling through droplets. The collision efficiency which, for water droplets striking ice, may be considered to be the total collection efficiency, is represented in Figure 6.12 as a function of droplet radius for various disk diameters. Fletcher points out, however, that there is no general agreement as to the correctness of these values.

A laboratory experiment on the collection of cloud ice crystals by a suspended ice sphere was performed by Hosler and Hallgren (1961). The ice crystals were driven past the ice sphere by a fan. The collection efficiency was determined by the ratio of the number of crystals collected to the number in the path of the collector. Spheres initially 127 and 360 microns in diameter were used, but the area presented to the flow by each increased because the growth was in the form of an aggregate of low density. The collection efficiency for prisms seemed to increase with decreasing temperature, but for plates it decreased. The results are shown in Table 6.3.

Growth to Precipitation in Warm Clouds

It is a simple matter to compute the growth of a drop which by some process or chance occurrence has become much larger than the average of the droplet population in a particular cloud. This situation can arise from the presence of giants in the nucleus population. Mason (1952) took the data of Frith (1951) for stratified clouds in which about one droplet in 5000 had a radius of 20 microns as contrasted with a volume-mean radius of 6 microns, and from "random walk" arguments computed that with a supersaturation of 0.05 per cent

TABLE 6.3

COLLECTION EFFICIENCY AND DENSITY OF COLLECTED AGGREGATE
FOR SMALL CRYSTALS ON ICE SPHERES

(After Hosler and Hallgren [1961])

TEMPERATURE (° C)		ICE SPHERES			
		127 μ dia.		360 μ dia.	
		Efficiency	Density	Efficiency	Density
Prisms 8–10μ	− 6.............	.044	.044	.057	.056
	− 8.............	.057	.029	.100	.043
	−10.............	.094	....	.133	....
Plates 8–18μ	−11.............	.104	.021	.210	.030
	−16.............	.068	.013	.104	.022
	−20.............	.034	.008	.081	.016
	−24.............	.028	.009	.058	.014

droplets grown on 10^{-12} to 10^{-14} g NaCl would have to remain in the cloud six to nine times longer to grow to 20 μ than to reach 6 μ. Whether one considers long-lived droplets—1 to 2 hrs in Mason's example—or giant nuclei, the end result in terms of vertical traverse in the cloud would be the same, although not the same in terms of time.

For a stratified cloud in which there is time for appreciable growth by vapor diffusion, the rate of growth can be calculated through three stages, one for $R < 20\ \mu$ when only diffusion growth enters, then for $20\ \mu < R < 60\ \mu$ when both diffusion and coalescence are considered, and finally with $R > 60\ \mu$ when only coalescence need be included. The growth rates are represented by equations (5.22) and (6.3) as

$$\frac{dR}{dt} = \frac{S-1}{R(a+b)} + \frac{u_T \chi E}{4 \rho_L} \qquad (6.19)$$

with each of the two terms included or omitted, as the case may be.

165

In terms of the height and time history of the drop in a cloud system containing an updraft U—or downdraft $-U$—one may write

$$z_2 - z_1 = \int_{t_1}^{t_2} (U - u_T)\,dt,\qquad(6.20)$$

and, with $dR(dR/dt)^{-1}$ substituted for dt,

$$z_2 - z_1 = \int_{R_1}^{R_2} (U - u_T)\frac{dR}{dR/dt}.\qquad(6.21)$$

It is also permissible to write $dR/dz = (dz/dt)^{-1}(dR/dt)$, and, since $dz/dt = U - u_T$,

$$\frac{dR}{dz} = \frac{u_T \chi E}{4\rho_L(U - u_T)},\qquad(6.22)$$

where it is assumed that the collected droplets are moving with the updraft at no terminal speed. A rearrangement gives

$$\frac{U - u_T}{Eu_T}\frac{dR}{dz} = \frac{\chi}{4\rho_L} = \left(\frac{U}{Eu_T} - \frac{1}{E}\right)\frac{dR}{dz},\qquad(6.23)$$

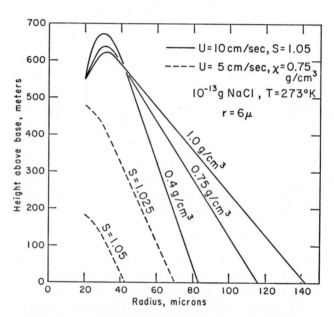

Fig. 6.13.—Plot of results of Mason (1952) on growth of droplet by coalescence in stratified cloud. Nucleus grows to droplet of $20\,\mu$ radius at given supersaturations, after which coalescence with droplets of $6\,\mu$ radius accounts for growth along the five curves for the specified conditions of updraft speed and liquid-water content.

or, in integrated form with a constant updraft,

$$U \int_{R_0}^{R} \frac{dR}{E u_T} - \int_{R_0}^{R} \frac{dR}{E} = \frac{1}{4 \rho_L} \int_{z_0}^{z} \chi dz . \qquad (6.24)$$

Since E and u_T are functions of R, the two integrals on the left are evaluated numerically. Figure 6.13 shows results calculated by Mason.

Langmuir (1948) and Ludlam (1951) used a modified form of equation (6.23) in a study of drop history in a convective cloud. A drop is considered to enter the cloud at the bottom and travel in the cloud until it grows to the point where it falls out of the bottom again. If E is a function of height only,

$$4 \rho_L \int_{R_0}^{R} \frac{(U - u_T)}{u_T} dR = \int_{z_0}^{z_0} E \, dz = 0 , \qquad (6.25)$$

$$4 \rho_L \int_{R_0}^{R} \frac{U}{u_T} dR - 4 \rho_L \int_{R_0}^{R} dR = 0 , \qquad (6.26)$$

and, finally,

$$R - R_0 = U \int_{R_0}^{R} \frac{dR}{u_T} . \qquad (6.27)$$

According to this expression the radius of a raindrop falling out of a cloud depends only on the updraft speed and the size the drop had when it entered the cloud.

For an R_0 of 40 μ Ludlam found the following values of R in terms of updraft U:

Updraft speed, U, m sec^{-1}........	1	2	3	4
Radius at emergence, mm........	0.3	0.9	1.6	2.4

Growth of Hailstones

An interesting case of growth by accretion of droplets is represented by the growth of hailstones. In this case, strong effects of heat exchange are present because a hailstone may or may not get rid of heat fast enough to freeze all the water it strikes. It is possible for a stone to grow while its outer surface is covered by a liquid layer or while liquid is being absorbed to form a "spongy" hailstone. Due to theoretical developments by Schumann (1938) and Ludlam (1958) it is possible to present a quantitative theory of hailstone growth.

As a first step, the transfer of heat to the environment by a spherical hailstone that remains at the melting temperature, 0° C, will be considered. With a ventilation factor a, the sensible heat transfer from the stone is given, according to (5.9), as

$$\frac{dQ_s}{dt} = 4 \pi R a K c (273 - T), \qquad (6.28)$$

167

or, with T in $°$ C and the specific heat c taken as 1, the expression is simply

$$\frac{dQ_s}{dt} = -4\pi R a K T. \tag{6.29}$$

The latent heat transfer to the environment is given, according to equation (5.7), as

$$\frac{dQ_L}{dt} = 4\pi R b L D \Delta \rho_w, \tag{6.30}$$

where b is the ventilation coefficient applicable to the vapor and $\Delta \rho_w$ is the difference in vapor density between the stone and the cloud air, taken as positive for evaporation. The combined heat transfer is

$$\frac{dQ_v}{dt} = 4\pi R(-aKT + bLD\Delta \rho_w), \tag{6.31}$$

which for a negative T ($°$ C) in a saturated environment will be positive.

For hailstone sizes the Reynolds number is always large ($Re \gg 100$), and under these conditions a and b are essentially equal, or about $0.3\, Re^{1/2}$. Then

$$\frac{dQ_v}{dt} = 4\pi R a(-KT + LD\Delta \rho_w). \tag{6.32}$$

Next we consider the transfer of latent heat of fusion as collected undercooled droplets are frozen to the stone. This heat is given by the mass accreted —in the case of 100 per cent collection efficiency, by the volume of the cylinder swept out times the liquid water content—multiplied by the latent heat of fusion, or

$$\frac{dQ_f}{dt} = \pi R^2 U_T \chi L_f. \tag{6.33}$$

The sensible heat exchanged in this collection is

$$\frac{dQ_s}{dt} = \pi R^2 U_T \chi T,$$

where T again is in $°$ C (negative for undercooled droplets). The combined heat from collected droplets is

$$\frac{dQ_c}{dt} = \pi R^2 U_T \chi (L_f + T). \tag{6.34}$$

The accretion of the undercooled droplets always transfers heat to the stone, because L_f is always larger than the negative T. Under this onslaught the stone may not be able to keep its outer part frozen, and so it becomes wet. The following situations are possible:

Dry stage: $\dfrac{dQ_v}{dt} > \dfrac{dQ_c}{dt}$.

Wet stage: $\dfrac{dQ_v}{dt} < \dfrac{dQ_c}{dt}$.

Equilibrium: $\dfrac{dQ_v}{dt} = \dfrac{dQ_c}{dt}$.

The growth rate of a hailstone by accretion is exactly as given in equation (6.3) for the case of the growth of a falling raindrop, or

$$\frac{dR}{dt} = \frac{U_T \chi}{4 \delta_r} \qquad (6.35)$$

for a collection efficiency of 1, where δ_r is the density of the coating of rime from the droplets. This expression would represent truly dry growth.

For the possibility of wet growth there are two terms in the equation, one for the effect of droplet accretion and the other for the exchange with the air and its vapor. The equilibrium may be written

$$\frac{dQ_f}{dt} + \frac{dQ_s}{dt} - \frac{dQ_v}{dt} = 0 , \qquad (6.36)$$

$$\pi R^2 U_T \chi L_f + \pi R^2 U_T \chi T - 4\pi R a (-KT + LD\Delta\rho_w) = 0 ,$$

which, when divided through by $4\pi R^2 L_f$, becomes

$$\frac{U_T \chi}{4} + \frac{U_T \chi T}{4 L_f} - \frac{a}{R} \cdot \frac{-KT + LD\Delta\rho_w}{L_f} = 0 . \qquad (6.37)$$

But the first term is $\delta_e (dR/dt)$ as in (6.35), where δ_e is the density of the ice forming the hailstone—usually about 0.9 g cm^{-3}—so the expression may be written

$$\frac{dR}{dt} = -\frac{U_T \chi}{4 \delta_e} \frac{T}{L_f} + \frac{a}{R \delta_e} \cdot \frac{-KT + LD\Delta\rho_w}{L_f} . \qquad (6.38)$$

In this equation $\Delta\rho_w$ may be taken as $\rho_{ws}(T_0) - \rho_{ws}(T)$—the saturation vapor density at the temperature of the hailstone minus that at the temperature of the ambient cloud—assuming equilibrium with respect to liquid water. With $a = 0.3 \, Re^{1/2}$ and $Re = 2RU_T\rho_a/\eta$, the factor $a/R\delta_e$ becomes

$$\frac{0.3}{\delta_e} \left(\frac{U_T \rho_a}{R\eta} \right)^{1/2} .$$

It is apparent that for a hailstone starting from some rudimentary ice pellet, the size as a function of time is inextricably tied in with height in the cloud and the cloud properties. The terminal velocity continuously changes with changing radius as well as with air density. The temperature, liquid-water content, vapor density, Reynolds number, thermal conductivity, diffusivity, and latent heats also are changing. The integration is obviously a problem for a modern computer. One needs a sounding or at least a computation of the saturation-adiabatic values of T, ρ_w, and χ as functions of pressure height. Updraft or downdraft speeds must be known or estimated, as well as the horizontal currents which may carry a hailstone from one part of the cloud to another or even out of the cloud. Das (1962) has shown that the horizontal transports can be important. Data on terminal velocities are available (Bilham and Relf, 1937; List, 1959a, 1961).

The possibility that the unfrozen water in wet growth remains as a thickening layer on the outside of the stone needs to be considered, for if any of this water is shed the whole balance will be affected. List (1959b) has found from extensive investigations in the hail tunnel on the Weissfluhjoch in Switzerland that hail in wet growth is characteristically spongy, taking up the water through intercrystalline spaces inside the stone, so that shedding of water is not observed.

Equation (6.38) may be used for determining the rate of decrease in size by melting in a warm environment, if the assumption that the water is shed as rapidly as it melts is valid. Collected water, unable to freeze, would also be considered as shed. In most cases of a stone's falling below the melting level it is probable that newly collected water cannot be held in a spongy stone, since the available storage space is decreasing while the amount to be stored is increasing. In a growing stone the available spongy space is always increasing.

REFERENCES

ACKERMAN, B. (1959), *J. Met.*, **16**, 191.

———— (1963), *J. Atmos. Sci.*, **20**, 288.

BARRETT, E. W. (1960), Instrument Society of America, Conference Preprint, No. 5-SF60 (Distributed by the Society at 313 Sixth Ave., Pittsburgh 22, Pa.).

BATTAN, L. J. (1953), *J. Met.*, **10**, 311.

BATTAN, L. J., and REITAN, C. H. (1957), in *Artificial Stimulation of Rain*, ed. H. WEICKMANN and W. SMITH (New York: Pergamon Press), p. 184.

BILHAM, E. G., and RELF, E. F. (1937), *Quart. J. Roy. Met. Soc.*, **63**, 149.

BRAHAM, R. R., JR. (1963a), *J. Appl. Met.*, **2**, 498.

———— (1963b), *Int'l. Assn. Met. Atmos. Phys. Proc. 13th General Assembly*, p. 143.

———— (1964), *J. Atmos. Sci.*, **21**, 640.

BRAHAM, R. R., JR., BATTAN, L. J., and BYERS, H. R. (1957), *Amer. Met. Soc., Met. Monogr.*, **2**, No. 11, 47.

BROWN, E. N. (1961), *J. Met.*, **18**, 815.

BROWN, E. N., and BRAHAM, R. R., JR. (1959), *J. Met.*, **16**, 609.

BROWN, E. N., and WILLETT, J. H. (1955), *Bull. Amer. Met. Soc.*, **36**, 123.

BYERS, H. R. (1959), in *The Atmosphere and the Sea in Motion* (New York: Rockefeller Institute Press), p. 400.

DAS, P. (1962), *J. Atmos. Sci.*, **19**, 407.

DRAGINIS, M. (1958), *J. Met.*, **15**, 481.

FLETCHER, N. H. (1962), *The Physics of Rainclouds* (Cambridge: Cambridge University Press), p. 281.

FRITH, R. (1951), *Quart. J. Roy. Met. Soc.*, **77**, 441.

GOYER, G. G., McDONALD, J. E., BAER, F., and BRAHAM, R. R., JR. (1960), *J. Met.*, **17**, 442.

GUNN, K. L. S., and HITSCHFELD, W. (1951), *J. Met.*, **8**, 7.

GUNN, R., and KINZER, G. D. (1949), *J. Met.*, **6**, 243.

HOCKING, L. M. (1959), *Quart. J. Roy. Met. Soc.*, **85**, 44.

HOSLER, C. L., and HALLGREN, R. E. (1961), *Nubila*, **4**, No. 1, 13.

KINZER, G. D., and COBB, W. E. (1958), *J. Met.*, **15**, 138.

KRASNOGORSKAYA, N. V. (1963), *Proc. Internat. Conf. Atmospheric and Space Electricity* (Amsterdam: Elsevier Publishing Co., 1964).

LAMB, H. (1932), *Hydrodynamics* (Cambridge: Cambridge University Press), p. 602.

LANGLEBEN, M. P. (1954), *Quart. J. Roy. Met. Soc.*, **80**, 174.

LANGMUIR, I. (1948), *J. Met.*, **5**, 175.

LANGMUIR, I., and BOLDGETT, K. B. (1946), Army Air Forces Tech. Rep. No. 5418, Washington.

LEVIN, L. M. (1954), *Dokl. Akad. Nauk, USSR*, **94**, 467.

LIST, R. (1959a), *Zeitschr. angew. Math. Phys.*, **10**, 143.

———— (1959b), *Helvetica Physica Acta*, **32**, 293.

———— (1961), *Nubila*, **4**, No. 1, 29.

LUDLAM, F. H. (1951), *Quart. J. Roy. Met. Soc.*, **77**, 402.

———— (1958), *Nubila*, **1**, No. 1, p. 1.

MACKLIN, W. C., and LUDLAM, F. H. (1961), *Quart. J. Roy. Met. Soc.*, **87**, 72.

MAGONO, C. (1953), *Sci. Rep. Yokohama Nat'l. Univ.*, Sec. 1, No. 2, p. 18.

———— (1954), *ibid.*, Sec. 1, No. 3, p. 33.

MASON, B. J. (1952), *Quart. J. Roy. Met. Soc.*, **78**, 275.

———— (1957), *The Physics of Clouds* (New York: Oxford University Press), pp. 422–24.

NAKAYA, U., and TERADA, T. (1934), *J. Fac. Sci. Hokkaido Univ.*, Ser. 2, 3, p. 1.

NEEL, C. B. (1955), NACA Res. Mem. No. A54123.

NEEL, C. B., and STEINMETZ, C. P. (1952), NACA Tech. Note No. 2615.

NEIBURGER, M. (1949), *J. Met.*, **6**, 98.

NIKANDRA, G. T., and KHIMACH, M. A. (1960), *Glavnaia Geofiz. Obs., Trudi*, No. 102, p. 50.

OSEEN, C. W. (1910), *Ark. Math. Astron. Fys.*, **6**, No. 29.

PEDERSEN, K., and TODSEN, M. (1960), *Norske Videnskaps-Akad., Geofys. Pub.*, **21**, No. 7.

PHILLIPS, B. B., and KINZER, G. D. (1958), *J. Met.*, **15**, 369.

RANZ, W. E., and WONG, J. B. (1952), *Industr. Eng. Chem.* (Industr.), **44**, 1371.

SCHOTLAND, R. M. (1957), *J. Met.*, **14**, 381.

SCHUMANN, T. E. W. (1938), *Quart. J. Roy. Met. Soc.*, **64**, 3.

SHAFRIR, U., and NEIBURGER, M. (1963), *J. Geophys. Res.*, **68**, 4141.

SINGLETON, F., and SMITH, D. J. (1960), *Quart. J. Roy. Met. Soc.*, **86**, 454.

SQUIRES, P. (1958), *Tellus*, **10**, 372.

TELFORD, J. W., THORNDIKE, N. S., and BOWEN, E. G. (1955), *Quart. J. Roy. Met. Soc.*, **81**, 241.

TWOMEY, S. (1956), *Tellus*, **8**, 445.

WARNER, J. (1955), *Tellus*, **7**, 449.

WARNER, J., and NEWNHAM, T. D. (1952), *Quart. J. Roy. Met. Soc.*, **78**, 46.

7
CLOUD DYNAMICS

In the study of cloud physics the microphysical processes occurring at size scales between molecules and large precipitation particles are emphasized. However, it should be noted that these processes would not come about unless the air circulations were favorable for the development of clouds and for producing the kind of cloud environment in which the microphysical events could proceed. Large sections of the atmosphere are without cloud and thus carry water only in the vapor form, even though the air contains an abundance of aerosols with which the vapor could react if the environmental saturation ratio were favorable. Saturation and supersaturation usually are reached by an adiabatic process in ascending air, as described in chapter 1. Radiation cooling to the dew point is sometimes observed in layers that are already near saturation, while at the surface fogs may form by contact with cold water or land.

The thermodynamics of air parcels as treated in chapter 1 provides a simplified description of the effects of ascent and descent in clouds, especially those of the convective type, but no model of a cloud—its size, shape, mass transport of air and water, relation to properties and air motions of its environment, rate of growth or dissipation, life cycle, etc.—can be obtained from this type of treatment.

With the development of radar and satellite cloud photography the visualization of cloud patterns on large and small scales has been greatly enhanced. Many patterns follow those found in laboratory models.

Bénard Cells

Numerous observers have noted that in unstable fluid layers convective motions occur in regular patterns called Bénard cells after the physicist who, in 1900, first described them in detail. Brunt (1951) outlined a simple home ex-

173

periment to demonstrate the effect. Inexpensive gilt paint consisting of gold-colored flakes suspended in a volatile fluid is poured into a shallow layer in a dish. The evaporational cooling of the surface of the fluid produces instability. The flakes tend to arrange themselves parallel to the motion so that when viewed from above they are less prominent in regions of ascent and descent than in regions of horizontal flow. A spacing of relatively clear and opaque areas outlines the cells, suggesting in each a core of ascending fluid surrounded by a ring of horizontal motion and an outer zone of descent. The remarkable regularity of repetition of the pattern, as shown in Plate 7.1, is the feature that attracts attention. A vertical slice through a simple Bénard cell showing the circulation is represented in Figure 7.1. Similar effects are seen in a variety of fluids cooled from above, and in molten metals. A cover over the liquid overcomes a

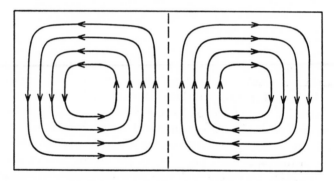

Fig. 7.1.—Circulation in a Bénard cell seen in vertical cross-section. The cell may be visualized as consisting of an ascending core or as having descent in the center surrounded by a ring of ascent, such as might be represented by adding a symmetric half-cell to the right or left of the sketch.

tendency for too violent, chaotic motions by reducing the rate of evaporation and cooling.

That the Bénard type of motion occurs in the atmosphere is suggested by satellite or high-altitude photographs of cloud patterns. In sections of the atmosphere where relatively shallow unstable layers with light winds occur over uniform surfaces, these features are common. Tropical and subtropical ocean areas are well suited to this type of cloud behavior. The photograph in Plate 7.3 is an example.

The effects of shear on the cells can be demonstrated in the laboratory by steadily moving across the liquid a glass plate that is in contact with the surface of the liquid. The shear thus produced is analogous to a change of wind with height in the atmosphere. The shear causes the cells to be distorted into elongated parallel rolls rotating in opposite directions. Shearing effects are shown in Plate 7.2 from laboratory models and from satellite photographs in Plate 7.4.

174

"Cloud streets" in which systems of convective clouds extend in lines over many miles with clear spaces between the lines have been observed by many persons flying at high altitudes over the oceans. One imagines them as representing Bénard types under shear. In all cases, with or without shear, the regular pattern occurs only when there is a well-defined upper limit to the convection. That is why these forms are found in the trade-wind region where the well-known trade-wind inversion, or stable layer, is present as a limit to the ordinary trade-wind cumulus in which these patterns are exhibited.

In violent convection, such as in thunderstorms, regular patterns of cells are not found. One reason is that such convection is not restricted by a well-defined upper boundary as is the case in true Bénard cells. It is observed that a system of thunderstorm cells so disturbs its local environment that regular patterns of clouds are destroyed.

Effects of the Turbulent Boundary Layer

Recent advances in the study of the turbulent boundary layer have led to a recognition that an inherent instability related to the shear flow in that layer can produce patterns of clouds. "Large eddies" having a scale quite distinct from the general spectrum of turbulence, but drawing their energy from the mean shear flow and expending it as eddy viscosity in the smaller scale, have been noted by several investigators and interpreted by Townsend (1956). In a series of fluid experiments in a rotating tank, Faller (1963) noted a banded appearance in the dye patterns in the bottom of the tank. Subsequently (1964), he interpreted these as related to the type of eddy structure that produces "cloud streets."

The Faller bands were found to originate at very nearly a constant value of the Reynolds number defined as $Re = VD/\nu$, where V is the tangential (zonal) speed above the boundary layer, ν is the kinematic viscosity, and $D = (\nu/\Omega)^{1/2}$ is the characteristic depth of the Ekman boundary layer.[1] The rotation rate is Ω. Thus the critical value of the Reynolds number is $Re_c = V/(\Omega\nu)^{1/2}$. Variations in the critical value were primarily caused by curvature of the flow in the tank, so the computed values of Re_c were extrapolated to zero curvature to produce a critical number for linear Ekman flow of $Re_c \approx 125$. The average spacing of the bands was found to be $L \approx 11\ D$.

The type of instability producing the bands has been given the name "inflectional instability," since the critical Reynolds number appears to be a function of the "degree of inflection" in the vertical profile of the flow. The inflection occurs above D, but its value is not rigorously defined. The L seems to depend inversely on the degree of inflection. The large eddies associated with the

[1] The flow in the boundary layer is that given by the Ekman spiral, defined in elementary textbooks of meteorology or oceanography. The factor D is essentially the height of the maximum in the Ekman wind profile.

175

bands have horizontal axes of rotation located above the Eckman layer, and they rotate in alternate directions to produce alternate bands of ascent and descent. The pattern in vertical cross-section, from a sketch by Townsend modified to make it appear more regular, is shown in Figure 7.2.

In the atmospheric case V is the geostrophic flow above the boundary layer, and the horizontal flow v within the boundary layer is a function of the geostrophic wind, the local earth rotation Ω, the turbulent viscosity v_t, and height

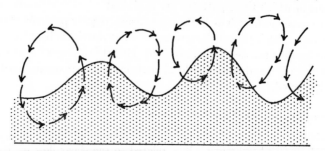

Fig. 7.2.—Development of large eddies (arrows) from the top of a perturbed boundary layer (shaded), as described by Townsend (1956).

z. This combination of variables permits the formulation of a non-dimensional expression

$$\frac{v}{V} = f\left[\frac{V}{(\Omega v)^{1/2}}, \frac{z}{(v/\Omega)^{1/2}}\right], \tag{7.1}$$

or, from the previous notation,

$$v = f_2(Re, z). \tag{7.2}$$

Since the turbulent eddy viscosity v_t is proportional to v^2, we may write for an average turbulent eddy viscosity,

$$v_t = \frac{v^2}{\Omega}[K(z_t)]^2, \tag{7.3}$$

where z_t is a non-dimensional vertical coordinate defined as $z_t = z\Omega/v$, an arbitrary but fixed scale height near the top of the boundary layer. For a given z_t it is seen that, with this definition inserted in (7.3), K is a constant. We now have a turbulent Reynolds number

$$Re_t = \frac{V}{(\Omega v_t)^{1/2}} = \frac{V}{\left\{\Omega \frac{v^2}{\Omega}[K(z_t)]^2\right\}^{1/2}} = \frac{V}{vK(z_t)}, \tag{7.4}$$

or, for a given z_t, $Re = 1/K$, and therefore is constant. The characteristic turbulent depth becomes

$$D_t = \left(\frac{v_t}{\Omega}\right)^{1/2} = \frac{v}{\Omega}K, \tag{7.5}$$

thus dependent on velocity.

176

In the atmospheric boundary layer Re is between 500 and 1000; D_t is of the order of 200 and 300 m. With $Re \equiv 750$ we find $K = 1.33 \times 10^{-3}$. With a wind of 10 m sec^{-1} and $\Omega \equiv 0.5 \times 10^{-4}$ sec^{-1}, it is found that the depth is

$$D_t = \frac{1.33 \times 10^{-2}}{0.5 \times 10^{-4}} = 266 \text{ m} ,$$

showing that the values are reasonable.

From a numerical integration of a relation derived from the Navier-Stokes equation, Faller determined that the vortex structure extends with significant amplitude to the height of $4D$. We have, therefore, a height of 800 to 1200 m, or 1064 m in the specific example just taken. For "wavelength" $L = 11$, across the band structure, $D_t = 2200$ to 3300 m or 2926 m in the specific example. The theory indicates that the axes of the large eddies should have an orientation 10 to 15 degrees to the left of the geostrophic wind. Since ν_t depends on the temperature lapse rate, the various quantities likewise will change with the lapse rate. The Ekman boundary theory is applicable to lapse rates near the dry adiabatic.

The significance of this study is that it indicates that cloud bands can develop from the Townsend large eddies generated from the Ekman boundary layer. If there is no excessive heating, these will transport all the heat necessary to be exchanged, and the pattern will be maintained. The stronger the wind the more dominant will be these large eddies; thus an explanation of the strong tendency for bands in hurricanes is suggested. In the bands themselves cumulus clouds with a spacing along the band characteristic of thermal convection may be found. In general, until a strong thermal convection develops, the shear-flow processes prevail in determining the pattern.

The Bubble Theory

For small cumulus clouds without shear and for dry thermals, the bubble model formally introduced by Scorer and Ludlam (1953) is widely used and accepted. A puff of buoyant air moves upward, spreading vertically and laterally in such a way as to move through a cone-shaped volume of air space. The model is often demonstrated by the reverse process of dropping a negatively buoyant slurry or salt solution into water and noting its spread and perturbation forms. It resembles in the reverse direction the rising top of a cumulus cloud. The circulation is like that of a vortex ring with ascent in the center and descent on the periphery and with new fluid brought into the wake.

A simple mathematical representation of the model using non-dimensional parameters has been given by Scorer (1958). To express the vertical velocity w a numerical constant C is introduced, and it is possible to write

$$w = C(g\bar{B}r)^{1/2} ,$$

177

where g is the acceleration of gravity, $\bar{B} = \overline{\Delta\rho/\rho}$, an average buoyancy term, and r is the radius of the bubble. For the vertical dimension, the preservation of shape requires that $z = nr$, where n is constant for a given bubble, and the volume is given by $V = mr^3$. In this type of model the buoyancy must diminish proportionately to the increase in volume, such that $V\bar{B} = V_1\bar{B}_1$, or

$$\bar{B} = V_1\bar{B}_1/V = r_1^3\bar{B}_1/r^3,$$

so

$$w = C(gr_1^3\bar{B}_1 r)^{1/2}r^{-3/2} = C(gr_1^3\bar{B}_1)^{1/2}r^{-1} = nC(gr_1^3\bar{B}_1)^{1/2}z^{-1} = dz/dt$$

and

$$z\,dz = nC(gr_1^3\bar{B}_1)^{1/2}dt$$

which, integrated between 0 and z, 0 and t becomes

$$z^2 = 2nC(gr_1^3\bar{B}_1)^{1/2}t.$$

We introduce the quantity $\beta = gV_1\bar{B}_1$ and find also that

$$\beta = gmr_1^3\bar{B}_1, \quad \beta^{1/2} = (gr_1^3\bar{B}_1)^{1/2}m^{1/2}, \quad \frac{\beta^{1/2}}{m^{1/2}} = (gr_1^3\bar{B}_1)^{1/2}; \quad z^2 = \frac{2nC\beta^{1/2}}{m^{1/2}}t.$$

Since n, m, C, and β are constant for a given thermal, we see that z^2 is linear in t for each thermal, or $z^2 = t/k$, and, furthermore, that $1/k\beta^{1/2} = 2nC/m^{1/2} =$ const. This constant is the slope of a line representing pairs of values of k^{-1} and $\beta^{1/2}$. Scorer indicates a value of $m \simeq 3$. (Note that for a sphere $m \simeq 4$.)

The outstanding feature of the bubble model is that it is *shape preserving*. From this simple model a dozen or more dynamic calculations by various authors appeared in the literature in the first decade of the theory's existence. Expressions involving an equation of motion, an equation of mass transport, and an equation of heat transport are usually invoked with eddy viscosity and diffusivity included. With the help of numerical integration a number of variations of the problem can be solved. Notable contributions have been made by Batchelor (1954), Kuo (1961), Levine (1959), Lilly (1962, 1964), Malkus and Witt (1959), Morton (1960), Morton, Taylor, and Turner (1956), Ogura (1962), Ogura and Charney (1962), Richards (1963), Saunders (1961, 1962), Scorer (1957), Turner (1957, 1960), and Woodward (1959). The solutions produce a circulation upward in the center and downward around the periphery, a circulation resembling a vortex ring, that is, with rotation around an axis forming an approximately circular horizontal ring.

Circulations in Convective Clouds

The bubble model appears to be suitable for thermally driven small cumulus clouds, at least in a crude sense, but for larger longer sustained convective clouds a jet model is a closer approximation. Perhaps the turrets or small towers of large cumulonimbus systems behave as bubbles.

Time-lapse motion pictures of clouds show that distinct thermal types seldom exist. In the first place, except for sun-drenched mountains and great fires, the effects of small local heat sources in cloud formation are so slight as to be totally obscured. Small cumuli more often represent wave crests in the horizontal air motions or are derived as large eddies from the turbulent boundary layer. The clouds are transitory features through which the air is flowing, producing condensation predominantly on the upwind side, dissipating with descent on the downwind side. The smallest type of cumulus and the air flow

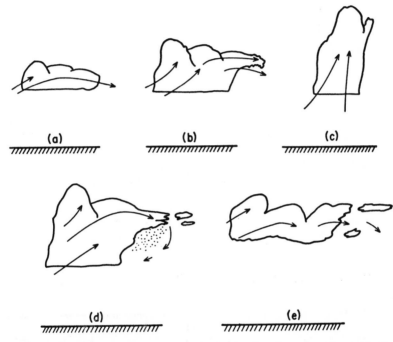

Fig. 7.3.—Some typical forms of cumulus clouds related to air motion. Clouds (*a*), (*b*), (*d*), (*e*) represent stages in development and dissipation with vertical shear; (*c*) represents conditions in relatively calm air. (From Byers, 1963.)

around it is represented in Figure 7.3*a*. As the buoyancy effect becomes stronger, a tower or turret is forced upward in a nearly vertical manner either so rapidly as to be negligibly tilted by the wind shear or with strong enough upward motion to make headway against the wind shear and to distort the wind pattern around it (Newton and Newton, 1959). But the circulation and the form of the cloud develop as in Figure 7.3*b*. The cloud has a hard side in the upshear direction and a soft side in the downshear direction. Some clouds in this condition, while still quite small, develop drizzle drops by the time the air reaches the soft side. The downward movement there circulates toward the lower part of the

179

soft side and the cloud dissipates while losing its bottom. Figures 7.3*d* and 7.3*e* show later stages of the same type of cloud in which it is able to grow by producing one or more new turrets on its upshear side despite erosion of the downshear base. In calm air the development might be as shown in Figure 7.3*c*, resembling bubble conditions, at least at the top.

Records of a flight through a cloud having a distinct hard and soft side are shown in Figure 7.4. In this case the shear was of the order of 3 m sec⁻¹ per km.

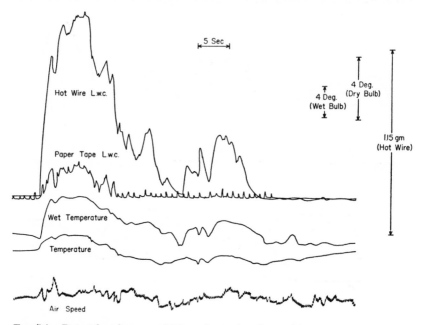

Fig. 7.4.—Data taken from a multichannel recorder of quantities measured in a flight at about 2700 m through a cumulus cloud in southern Missouri on June 7, 1963. The "hard" and "soft" sides, left and right, respectively, are clearly contrasted. Two types of liquid–water-content meters, the hot wire and the paper tape, were used. Measurement scales for some of the recordings are given on the right. Flight speed is about 75 m sec⁻¹. (From Byers, 1963.)

A "starting plume" model studied by Turner (1962) is suggestive of the growth of turrets on large clouds. This model has features of the bubble in that it takes into account mixing at the top, and the shape of the upper part is preserved, but it is fed by a steady jet from below. The jet, which forms the "stem" of the turret, eventually becomes non-buoyant by mixing with the environment air. Subsequently, as shown by Turner's model experiments (Turner, 1964), the plume shape changes with downflow around the edges until the plume falls or collapses back down around the jet. This sequence is more nearly like that occurring in natural cloud turrets than is found from the bubble model.

180

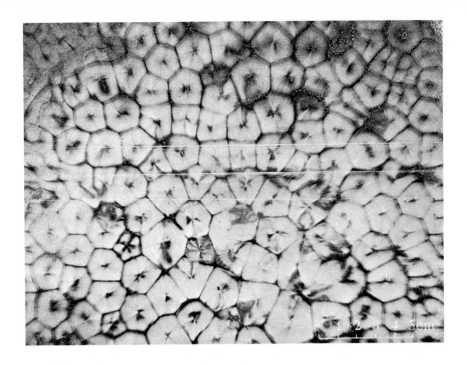

PLATE 7.1.—Bénard cells produced in a thin layer of fluid in the laboratory. (From K. Chandra, 1938.)

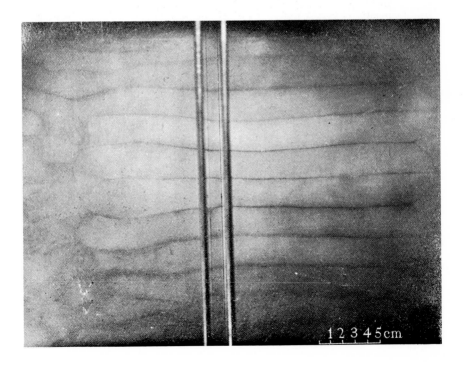

PLATE 7.2.—Effects of shear on cells in the laboratory model (from Chandra, 1938)

PLATE 7.3.—Satellite picture of convective cloud cells of Bénard type over land along the coast of North Carolina. The picture, which covers a 70-n-mi-square, shows Cape Fear at lower right; New River Inlet (Camp Le Jeune) are in the cloud-free area above and to the left of center. North is to the left. Date: November 25, 1960. Courtesy V. J. Oliver, U.S. Weather Bureau, National Satellite Center.

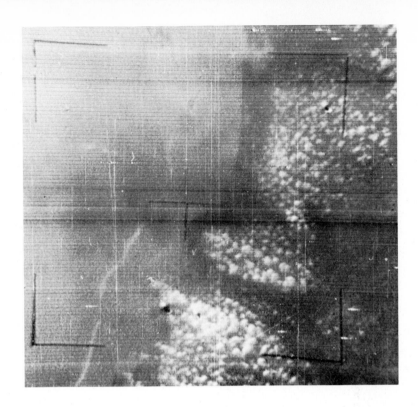

PLATE 7.4.—Effects of shear on natural atmosphere as pictured by satellite. The upper picture covers a 70-n-mi-square along the Gulf Coast from Mobile Bay westward toward the Mississippi Delta. West is at the top. Date: May 15, 1960. In the lower picture, the view covers a large area centered on Japan with approximate southeast at the top. Cold air from the Asiatic continent crossing the Sea of Japan forms clouds in "streets," lower right, then a solid mass over Japan, with a large, clear area on the leeward side. Cloud bands form again over the Pacific in the upper part of the picture. Date: January 19, 1964. Courtesy V. J. Oliver, U.S. Weather Bureau, National Satellite Center.

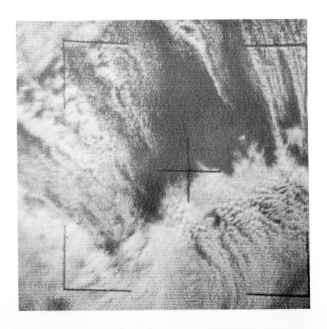

The Steady-State Jet Model

Cumulonimbus clouds contain steady updrafts and downdrafts (Byers and Braham, 1949) which have jetlike profiles. Theoretical treatments are simplified if a steady-state condition can be introduced, and the steady-state updraft has been most commonly treated. This simplification does not permit the transition to a downdraft, but once attained it can in turn be used as an initial state from which the triggering of the downdraft and the production of negative buoyancy can be developed.

Squires and Turner (1962) have considered a steady-state turbulent, buoyant, condensing plume or jet in which the problem is specified by the radius and velocity of the updraft at the cloud base, given an atmospheric sounding. Four basic expressions are used: an equation of continuity of mass flux, an equation for the increase of momentum flux due to buoyancy alone, a buoyancy expression, and an equation for the contribution of water substance.

The mass flux M (g sec^{-1}) is given by $M = b^2 w \rho$, where b is the radius of the plume, w the vertical velocity, and ρ the air density. The change of this quantity with height is

$$d(b^2 u \rho)/dz = 2b \, aw \rho_0 \,,$$

where a is an entrainment constant and ρ_0 is the density of the environment air. This expression indicates that the mass flux is increased at a rate proportional to the vertical velocity and the radius of the updraft. From jet experiments in the laboratory with cylindrical or "top-hat" velocity profiles, Morton (1960) gives the value $a = 0.116$ which appears to be good within ± 15 per cent.

One can write

$$\frac{1}{M}\frac{dM}{dz} = \frac{2b \, aw \, \rho_0}{b^2 w \, \rho} \approx \frac{2a}{b} \,,$$

where the last approximation means that if $\rho_0/\rho = 1$, the fractional rate of change of mass flux in a plume of given size is proportional to the rate of entrainment, and in the usual case of small density differences it is important only that a/b be fixed.

For the change in momentum flux due to buoyancy alone, the expression is

$$\frac{d}{dz}(Mw) = M\frac{dw}{dz} = M\frac{dw}{dt}\frac{dt}{dz} = \frac{M}{w}\frac{dw}{dt} \,,$$

and, considering that dw/dt is the buoyancy acceleration, we have

$$\frac{d}{dz}(Mw) = \frac{b^2 w \rho}{w} \, g \, \frac{(\rho - \rho_0 - \rho \sigma)}{\rho} = b^2 g (\rho - \rho_0 - \rho \sigma),$$

where σ is the liquid-water content in grams per gram of air. Thus the effect of the density of the condensed water on the buoyancy is considered.

181

The thermodynamic and mass flux (entrainment) contribution to lapse rate and buoyancy is derived from the classical saturation-adiabatic equation with neglect of the heat capacity of the condensed water. The temperature change in ascent is given by

$$\frac{dT'}{dz} = (T_0' - T')\frac{1}{M}\frac{dM}{dz} - \frac{\Gamma \rho_0}{\rho} - \frac{L}{M c_p}\left(\frac{dMq}{dz} - q_0\frac{dM}{dz}\right)$$
$$+ \frac{1}{\gamma}\left[\frac{d}{dz}(\delta T) + (\delta T - \delta T_0)\frac{1}{M}\frac{dM}{dz}\right],$$

where the prime refers to the virtual temperature, q is the mixing ratio inside and q_0 that outside the cloud, and the other symbols are as in chapter 1.

The third term on the right contains the contribution of water substance which may be written

$$\frac{dM\sigma}{dz} = q_0\frac{dM}{dz} - \frac{d}{dz}(Mq).$$

The Clapeyron-Clausius equation is written in the form

$$\frac{dT}{dz} = \frac{RT^2}{\epsilon L e}\frac{de}{dz},$$

where e is vapor pressure and $\epsilon = \rho_w/\rho$. The virtual temperature is given by $T' = T(1 + \lambda q)$, where $\lambda = (1/\epsilon) - 1$. Algebraic arrangement gives

$$\frac{dM}{dz} = \frac{2\alpha p^{1/2}}{R^{1/2}T_0'}(MwT')^{1/2},$$

$$\frac{dw}{dz} = \frac{g}{w}\left(\frac{T'}{T_0'} - 1 - \sigma\right) - \frac{w}{M}\frac{dM}{dz}.$$

By setting $D = 1 + \lambda q + [(q\epsilon\lambda L)/RT]$, one finally obtains

$$\frac{dT'}{dz} = \left\{\frac{D}{M}\frac{dM}{dz}\left[(T_0' - T) - \frac{L}{c_p}(q - q_0) + \frac{(\delta T - \delta T_0)}{\gamma}\right]\right.$$
$$\left. - \frac{D\Gamma T'}{T_0'} - \frac{L}{c_p}\frac{gq}{RT_0'}(1 + \lambda q) + \frac{\lambda}{\gamma}\frac{g qT}{RT_0'}\right\},$$

all divided by

$$\left[D + \frac{L}{c_p}\frac{q\epsilon L}{RT^2} - \frac{\lambda q}{\gamma}\left(\frac{\epsilon L}{RT} + 1\right)\right],$$

and for the liquid-water distribution,

$$\frac{d\sigma}{dz} = (q_0 - q - \sigma)\frac{1}{M}\frac{dM}{dz} - \frac{q}{D}\left[\frac{\epsilon L}{RT^2}\frac{dT'}{dz} + \frac{g}{RT_0'}(1 + \lambda q)\right].$$

The governing equations are integrated with height as the independent variable, the boundary conditions being given by the properties of the updraft at the cloud base. The updraft air at the base is assumed to have the same virtual temperature as the environment. Given a sounding and the cloud-base level, the problem is then specified by the updraft at this level.

Thunderstorm Cells

The now-familiar picture of cells observed on the Thunderstorm Project (Byers and Braham, 1949) is shown in Figure 7.5 in the "mature" stage. In the multilevel flights through the thunderstorms it was found that the drafts were continuous at any given time from the low levels to the highest altitudes and were maintained as jets for some period of time. Usually they were sloping in the sense of the wind shear. Cell boundaries were easily identified both from past evolution and relative absence of vertical motion, even though enmeshed in the cloud system.

It was observed that the radar cloud associated with a single thunderstorm cell was about 6 to 10 km in diameter. However, measurements by airplane showed the most frequent width of the updraft and downdraft area to be about 1500 m each. If the shape of the drafts is circular, this indicates that the major vertical motion within the cell is limited to about one-seventh of the cross-sectional area of the cell.

From range-height indicating radar it was noted that the thunderstorm tops ascended in a series of steps, appearing as the growth of new turrets from the cloud top. Large clouds usually exhibited several turrets during their growth. In 32 cases studied the maximum height reached was correlated with the number of turrets formed by a coefficient of $+0.67 \pm 0.10$. Each successive turret was higher than the preceding one, and the mean lapse of time between successive turret peaks was 17.8 min. It is believed that each turret made it easier for the following ones by increasing the moisture content in the cloud-top environment air which might be entrained in further growth. Less heat of condensation is robbed from the new turret by entraining than would have been the case if it were standing alone in a dry environment. Each turret underwent a growth period followed by an interval of subsidence. The growth period averaged about 16 minutes in duration. The average vertical growth rate was about 6 m sec^{-1} and the subsidence rate was about 4 m sec^{-1}.

These observations show that, except for the turrets, the updraft is not surrounded by clear air. The turrets behave as starting plumes. One assumes that the turrets represent tops of small jetlike updrafts.

Stored Water and the Downdraft

The Squires and Turner jet model, like others of its kind, cannot produce a downdraft because the condensed water is not permitted to accumulate and fall. On the Thunderstorm Project it was reasoned qualitatively that either the weight or the drag of copious amounts of falling water carried the air downward

183

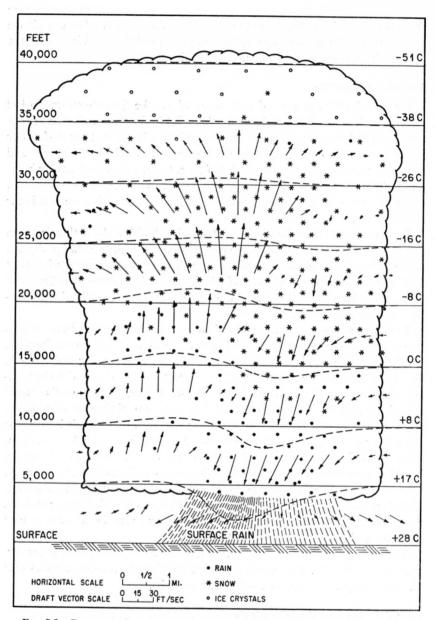

FIG. 7.5.—Representation of a thunderstorm cell in mature stage, including air motions (arrows on vector scale given in lower left), hydrometeors, and temperatures (dashed lines). (From Byers and Braham, 1949.)

until it achieved negative buoyancy. The effect of the dead weight of the water in a closed, adiabatically ascending parcel is easily determined (Saunders, 1957; Mason and Emig, 1961). The generating rate of liquid water has been deduced by Kessler (1959).

Das (1964) has considered the theoretical aspects of the effects of liquid water in initiating the downdraft. His model utilizes the classical approach similar to that of Squires and Turner but with quite different relations and conditions. The updraft part of the model is an elaboration of one introduced by Gutman (1957, 1961). As in similar treatments, the model is taken as one-dimensional with only implications concerning the horizontal dimensions. Three basic equations are used: (1) an expression relating vertical velocity to buoyancy acceleration and inflow; (2) an equation for the variation of the buoyancy with thermal stability and liquid-water content; and (3) an expression for condensed water distribution from generation in the updraft and removal by precipitation. All quantities were put in non-dimensional form.

The initial state is the distribution in a steady-state non-precipitating updraft. Then, with the terminal velocity—therefore drop size—taken as directly proportional to the liquid-water content, assuming that the number concentration of the drops is constant in each model, the water is allowed to redistribute itself. The result is an accumulation of water in the lower half of the cloud. The downward flux of water becomes greatest near the base; negative buoyancy soon characterizes the lower part of the cloud and the downdraft quickly develops there, accompanied by horizontal divergence below the cloud.

In spite of the simplicity of the model, it has shown considerable success in reproducing the general qualitative aspects of some of the observations of the thunderstorm downdraft.

REFERENCES

BATCHELOR, G. K. (1954), *Quart. J. Roy. Met. Soc.*, **80**, 339.

BÉNARD, H. (1900), *Rev. gén. de science pur et appl.*, **11**, 1261 and 1309.

BRUNT, D. (1951), in *Compendium of Meteorology*, ed. T. F. MALONE (Boston: American Meteorological Society), p. 1255.

BYERS, H. R. (1963), *Int'l. Assn. Met. Atmos. Phys., Proc. 13th Gen. Assbly.*, p. 85.

BYERS, H. R., and BRAHAM, R. R., JR. (1949), *The Thunderstorm* (Washington: U.S. Weather Bureau).

CHANDRA, K. (1938), *Proc. Roy. Soc. London, Ser. A*, **164**, 231.

DAS, P. (1964), *J. Atmos. Sci.*, **21**, 404.

FALLER, A. J. (1963), *J. Fluid Mech.*, **15**, 560.

——— (1964), paper presented at Nat'l. Conf. Phys. and Dyn. Clouds, Chicago, March 25, 1964 (Amer. Met. Soc.).

GUTMAN, L. N. (1957), *Dokl. Akad. Nauk, USSR*, **112**, 1033.

——— (1961), *Izvest. Akad. Nauk, USSR, Geophys. Ser.*, No. 7, p. 1040.

KESSLER, E, III (1959), *J. Met.*, **16**, 630.

KUO, H. L. (1961), *J. Fluid Mech.*, **10**, 611.

LEVINE, J. (1959), *J. Met.*, **16**, 653.

LILLY, D. K. (1962), *Tellus*, **14**, 148.

——— (1964), *J. Atmos. Sci.*, **21**, 83.

MALKUS, J. S., and WITT, G. (1959), in *The Atmosphere and the Sea in Motion* (New York: Rockefeller Institute Press), p. 425.

MASON, B. J., and EMIG, R. (1961), *Quart. J. Roy. Met. Soc.*, **87**, 212.

MORTON, B. R. (1960), *J. Fluid Mech.*, **9**, 107.

MORTON, B. R., TAYLOR, G. I., and TURNER, J. S. (1956), *Proc. Roy. Soc. London, Ser. A*, **234**, 1.

NEWTON, C. W., and NEWTON, H. R. (1959), *J. Met.*, **16**, 483.

OGURA, Y. (1962), *J. Atmos. Sci.*, **19**, 492.

OGURA, Y., and CHARNEY, J. G. (1962), *Proc. Internat. Symp. Weather Prediction* (Tokyo: Meteorological Society of Japan).

RICHARDS, J. M. (1963), *J. Atmos. Sci.*, **20**, 241.

SAUNDERS, P. M. (1957), *Quart. J. Roy. Met. Soc.*, **83**, 342.

——— (1961), *J. Met.*, **18**, 451.

——— (1962), *Tellus*, **14**, 177.

SCORER, R. S. (1957), *J. Fluid Mech.*, **2**, 583.

——— (1958), *Natural Aerodynamics* (London: Pergamon Press).

SCORER, R. S., and LUDLAM, F. H., (1953), *Quart. J. Roy. Met. Soc.*, **79**, 94.

SQUIRES, P., and TURNER, J. S. (1962), *Tellus*, **14**, 422.

TOWNSEND, A. A. (1956), *The Structure of Turbulent Shear Flow* (Cambridge: Cambridge University Press).

TURNER, J. S. (1957), *Proc. Roy. Soc. London, Ser. A*, **239**, 61.

——— (1960), *J. Fluid Mech.*, **7**, 419.

——— (1962), *ibid.*, **13**, 356.

——— (1964), paper presented at Amer. Met. Soc. Conf. Phys. and Dyn. of Clouds, Chicago, March, 1964.

WOODWARD, B. (1959), *Quart. J. Roy. Met. Soc.*, **86**, 144.

INDEX